DESPERATELY SEEKING SOLUTIONS

Helping students build problem-solving skills to meet life's many challenges

Kathy Paterson

Pembroke Publishers Limited

© 2009 Pembroke Publishers
538 Hood Road
Markham, Ontario, Canada L3R 3K9
www.pembrokepublishers.com

Distributed in the U.S. by Stenhouse Publishers
480 Congress Street
Portland, ME 04101
www.stenhouse.com

We acknowledge the financial support of the Government of Canada through the Book Publishing Industry Development Program (BPIDP) for our publishing activities.

We acknowledge the assistance of the Government of Ontario through the Ontario Media Development Corporation's Ontario Book Initiative.

Library and Archives Canada Cataloguing in Publication

Paterson, Kathy, 1943-
    Desperately seeking solutions : helping students build problem-solving skills to meet life's challenges / Kathy Paterson.

Includes index.
ISBN 978-1-55138-247-0

1. Problem solving — Study and teaching (Elementary). 2. Problem solving — Study and teaching (Secondary). I. Title.

BF723.P8P28 2009   370.15'24   C2009-903876-5

Editor: Kate Revington
Cover Design: John Zehethofer
Typesetting: Jay Tee Graphics Ltd.

Printed and bound in Canada
9 8 7 6 5 4 3 2 1

Mixed Sources
Product group from well-managed forests, and other controlled sources
www.fsc.org Cert no. SW-COC-002358
© 1996 Forest Stewardship Council
FSC

Contents

1 Approaching Problems with Care

A mother in her sixties shared with me her horror when her 32-year-old daughter, whom I'll call Amy, was stopped by police at a routine check stop. Amy was on her way to work as a waitress. She was unable to produce any identification, car registration, or car insurance documents. What problem-solving tactics did she employ? She called her mother on her cellphone, desperately asking what she should do.

In a similar tale of woe, a man in his early forties, whom I'll call John, accidentally broke an expensive company tool while at work on a job site. Panicked, he left the site and called his 65-year-old mother, desperately seeking a solution.

These incidents have nothing to do with kids, school, and teachers, you might say, but indeed they do. Neither the woman nor the man in question could problem-solve. Somehow Amy had got to the age of 32 and John to his early forties without ever learning how to get themselves out of an unexpected mess. All they could do was "call mom."

Whose responsibility was it to teach them how to problem-solve in any situation? I believe that responsibility falls equally to teachers and parents, and since kids often see more of their teachers than their parents, well, you can see where I'm going with this ...

In Support of Problem-Solving for Kids

A situation more home to teachers is the following. While on early morning supervision in the schoolyard, I noticed a little girl sitting on the grass sobbing.

"Oh my, Chelsea," I said, squatting down beside her. "What on earth is the matter?"

She gulped a breath then answered with a lengthy dissertation that went something like this. "... alarm clock ... mom mad ... late ... no breakfast ... couldn't find ... missed bus ... dad drove ... FORGOT MY REPORT!" (There was probably more that I can't recall now.) By the time she got to the punch line, she had exhausted herself by trying to talk and cry at the same time.

I assumed the role of calm mentor and said, "So, Chelsea, what do you think you could do to solve this problem?"

I'll never forget her response. She shouted at the top of her lungs — "CRY!"

In another scenario that the mother of a student shared with me, the mom heard a loud scream coming from her adolescent daughter's room. Fearing the worst, she hurried to her daughter. The girl, Tara, was literally pulling at her hair.

"What on earth ...?" the mom shouted.

"My cell! I've lost my cell and it has ALL my numbers on it!" Tara was beside herself.

"Okay," the mom said calmly, "Let's try to solve this problem. First …"

"If you tell me first not to panic," screamed the girl, "I'll just *die*. This is an emergency and I *want* to panic. I *have* to panic. I'm *panicking*!"

So Amy, John, Chelsea, and Tara, the turbulent teen, were all missing problem-solving skills. Each of them presented a common, badly chosen, and futile method of dealing with a problem. Amy and John cried for *help*, Chelsea accepted *defeat*, and Tara *panicked*. Their reactions were ineffective, inappropriate, and irrational. Perhaps if they had been trained to problem-solve and given experience in using good problem-solving techniques, they might have tried other more successful actions.

"Good" problem-solving techniques can be considered rational, so I will adopt the term "rational problem-solving," which refers to the combination of abilities to recognize, analyze, recall, visualize, weigh, compare, and predict. The American educator and philosopher John Dewey initially identified these. *Desperately Seeking Solutions* is based on the premise of rational problem-solving.

Yet so much has already been written about problem-solving that the need for this book may seem questionable. This book, however, deals with the timely topic mainly from the point of view of kids. How do we, as adults, teachers, friends, mentors, coaches, parents, specifically help kids become good problem-solvers in their everyday lives? How do we help them to make wise decisions, and if they don't, to live with the consequences?

I would argue that it is our responsibility to help our students and our children control their own destinies by learning how to solve problems competently. As Albert Ellis, an American psychologist and author, wrote: "The best years of your life are the ones in which you decide your problems are your own. You do not blame them on your mother, the economy, or the president. You realize that you control your own destiny." That is what we want for the children in our lives.

Human beings are meant to be problem-solvers, but some of us do it better than others. Because we have free will, we are continually faced with forks in the road, with distressing dilemmas, with difficult choices. Some people, perhaps as a result of immaturity, emotional distress, past history of ineffective problem-solving, or even physical illness, would rather not deal with problems at all. Others tend to see having a problem as a problem in itself and shy away from it. But most of us, would have it no other way. We accept problems as our own and look to the courage we will gain once they are solved. We understand that the ways we handle ourselves in moments of challenge and choice reveal who and what we are.

Consider a few personality types. We all know the weak help-me type; similarly, we know the stubborn I-can-do-it-alone type. A person of either type is not acting wisely. Depending totally on others or never on anyone else is not effective. People of both types may lack self-confidence and fail to use rational problem-solving techniques; instead, they adopt defences. In the first case, that defence is helplessness; in the second, it may be an unwillingness to accept personal fallibility. As educators and parents, we want children to grow up to be a functional combination of both these types and more. We want them to be rational problem-solvers, especially when no one is around to help.

Children, in fact, problem-solve all the time, without being aware of what they are doing. Only when they come fact to face with a new dilemma, a situation that is unexpected, even frightening, one requiring difficult decision making and action, do many become stumped. They lack the specific skills, the rational problem-solving skills, with which to react effectively in such a situation.

Rational problem-solving involves both critical and creative thinking, things that teachers can promote. Although kids may readily tap into both types of thinking on a regular basis, when faced with a tough problem, a roadblock to their smooth progression, they may forget how to think in either or both ways. So we teach them how. We provide concrete examples and help students work through them to logical conclusions. We talk about problem-solving, we model problem-solving, and we reinforce effective, rational problem-solving. We teach the vocabulary of rational problem-solving, and we teach the steps. We teach them how and why to evaluate the gravity of every problem.

We want students to be able to discern quickly the degree of gravity so that they can take appropriate measures. This step is important in effective problem-solving for, as you know, dealing with a broken nail, for example, requires much less mental fortitude than dealing with a broken promise. The latter requires specific rational problem-solving; the former does not.

All problem-solving begins with determining the gravity of a situation. Numerous adults have confided that they are weak when it comes to problem solving, to dealing effectively with stressful situations, to rising up and successfully meeting difficult challenges. It doesn't matter whether a difficulty is life threatening or simply annoying. They do not know how to respond.

Given that many adults find it hard to approach various situations, we can be certain that many children do, too. We can help. We can specifically teach the steps, approaches, and skills that enable effective rational problem-solving. We have to. Consider the following anecdote.

A man found a snake. Although it was frozen, he recognized it as a poisonous snake. He picked the poor thing up and took it home to revive it. He placed the snake in front of the fire to thaw. Once thawed, he bent down to give it a nice little saucer of milk. The snake lifted its head and bit the good man. As the man lay dying he asked the snake, "How could you do this to me after all I have done for you?" As the snake slid out the door, it turned to the good man and said, "Stop your whining. You knew I was a poisonous snake when you picked me up. What else did you expect me to do?!" (*Respect*, Aug. 1996)

The man in the anecdote failed to take into consideration the first step in problem solving: to consider the gravity of the situation. It appears he didn't use good, rational problem-solving skills, and the consequences were, unfortunately, disastrous. Perhaps the man was not blessed with a teacher or other significant adult who helped him become a rational problem-solver. Certainly, the role of the adult cannot be overlooked.

I realize the predicament this puts teachers in. You likely already feel like that you are in 10 places at once, trying to do 10 things at once, and do them all equally well. You likely feel overburdened by too full classes, too many students with special needs, too heavy a curriculum, too many bureaucratic demands, too many extra-curricular responsibilities, and not enough time. How can I possibly suggest adding even more to your teaching demands? I suggest it because I believe firmly that by teaching students to problem-solve on their own, teachers will be enabling them to do better in all the other subjects. I am suggesting that teaching effective problem-solving techniques should be an educational priority — perhaps the single most important skill we can develop in our charges. In a world that is rapidly changing daily, perhaps the best thing we can do to prepare young people for uncertain futures is to teach them to face and solve their own problems successfully, independently, and with self-confidence.

With this in mind, and with my sincere appreciation for the plight of over-worked teachers, I have included a 10-lesson mini-unit plan for teaching problem-solving to kids (see Appendix C). The plan comes with the lesson behavioral objectives, anticipatory sets, and student follow-ups — the three most difficult parts of lesson plans, as far as I'm concerned. Teachers will easily be able to fill in the rest according to class needs. For example, they may wish to use favorite stories or poems that fit with the stated objectives, or involve students in tasks or activities, such as role playing or collage making, which they have earlier found productive.

I hope you find this tool useful.

What Makes a Problem a Problem?

Problems are problems mainly because of two constructs: fear and deficiency. It is worthwhile to take a closer look at this idea since understanding what's behind a problem may help achieve a solution to it. The more people, and especially children, understand a paradigm, the better chance they have of facing it with confidence. We are all familiar with the idea of fear of the unknown; the more we can enlighten students about what is behind the scenes when a problem arises, the better equipped they will be to solve it.

Fear

Dostoevsky wrote that "taking a new step ... is what people fear most." If his observation is true, every time we face a problem, there is a degree of inherent fear. As teachers we need to be fully aware of this and to empathize with our students who may appear reluctant to address their problems.

Having said that, it is equally important to recognize that fear creates most problems. Fear of rejection, punishment, loneliness, sorrow, anxiety, failure — basically, the fear of either emotional or physical hurt. Consider the following:

- Problem — lost car keys: Fear of being late, losing job, being frowned upon, missing appointment
- Problem — assignment undone: Fear of teacher wrath, poor grades, parents' disappointment
- Problem — having to admit to an inappropriate act: Fear of rejection, disappointment of a loved one, punishment, personal self-disrespect
- Problem — stole an item from teacher's desk: Fear of punishment, parental involvement, and disappointment, embarrassment if peers find out

You can see how fear is behind almost everything that causes a problem. Even when we think our fear is for another person, as in the case of not wanting to tell a friend she has failed at something because we are afraid that she'll feel badly, the underlying fear is personal. We are afraid of our own feelings when

Teacher Tip

Present a problem. One example is that you have told a secret and your lack of trust has been discovered. As a class, discuss what hidden fears could be behind it. Doing this can serve to open kids' eyes to the fact that everyone experiences fear and deals with it often. They may find this activity empowering.

we inadvertently cause someone else to feel badly. It's a paradox — a curious "fear of fear" situation — that lurks behind most problems.

So, what do we do with this information when teaching students to problem-solve? We tell them to look for the little hidden fear behind the problem, and when they can identify it, they will be able to face the problem squarely and deal with it more effectively.

Deficiency

Although it is arguable that deficiency is just another fear, I have separated it because I have found that children are better able to understand it as a separate entity. *Deficiency* refers to the lack of something, primarily money or skill. Consider the problem of wanting to attend summer camp but being unable to afford it. You could argue that fear is at work — fear of being left out, fear of loss of experience, and so on, but the primary behind-the-scenes interference here is deficiency — lack of monetary resources. Consider the following:

- Problem — want to learn to play piano or to dance; Deficiency: money, current skills
- Problem — ripped good jacket; Deficiency: money to repair or replace

So, what is the point of this behind-the-scenes concept? A quick lesson in which students are encouraged to look for what is behind a problem can often give them a head start in finding a good solution. This is not an in-depth examination of causes. (In order to fully examine all the myriad causes of problems, another book would be necessary.) Instead, this approach is a brief investigation based on two constructs. It is meant to be self-empowering.

Role of the Adult in Teaching Kids to Problem-Solve

Why is it that some children, and adults, can solve problems no matter how difficult the situation, whereas others seem to give up immediately? There are many reasons for these opposing reactions, lack of training and practice in problem-solving being the most obvious. As teachers and also as parents, we can help children overcome these lacks.

We have a significant role to play in teaching the young how to face and deal effectively with problems, to problem-solve. Consider the young student, Chelsea, in the first section of this chapter. Without intervention by an adult, she could well grow up to be just like John and Amy, calling home to a parent the moment she's in trouble. Or consider the teenager Tara as an adult, screaming and panicking at a board meeting when she is faced with a differing opinion from a co-worker. These behaviors are not what we want for our young people, be they students or offspring. We want them to be independent learners, independent problem-solvers, and independent survivors. We want them to grow up and live happy, productive, self-disciplined, and self-confident lives. We want them to be rational problem-solvers.

Effects of adult actions on kids' ability to problem-solve

See Chapter 9 for an expansion on this theme.

Psychology has helped to identify a number of adult actions that may either encourage or inhibit problem-solving techniques in children. Unfortunately, some adult actions, although delivered with the best of intentions, may heighten the youngsters' difficulties or even create co-dependencies — those restrictive, unhealthy connections between two persons where each is overly dependent on the other. Other adult actions tend to have the opposite, more positive effect; they may well serve to increase the likelihood of kids' ability to problem-solve and thereby achieve success. The following list of behaviors provides an overview of interactions that we, as teachers, mentors, or parents, should either adopt to promote children's successes or avoid to prevent difficulties or dependencies.

What to Do

- **Set boundaries:** Let kids know what you expect of or from them not just when a problem arises, but all the time. (You will be home by 9 p.m. Or, you will be sitting in your desk when the last bell goes.) Then, when a problem arises, it's easy to remind them of these boundaries.
- **Deal with manipulation:** When a child attempts to manipulate you or a situation rather than facing a problem, step in and stop him. (A boy tells his teacher he was unable to do his homework because hockey practice took all his energy. He smiles sweetly and invites the teacher to attend his next game. The teacher points out that this is manipulation on his part and is unacceptable.)
- **Reinforce all solutions:** Whether the child makes a good or a bad choice, reinforce the fact that she attempted to solve the problem. ("I'm proud of you for making your own decision about what to bring to the track meet. It's unfortunate you didn't bring rain gear, but next time you'll make a better choice.")
- **Offer to help:** Offer support, with statements such as "How can I help you solve your problem?" (Perhaps a child desires to purchase a gift for a parent, but lacks enough money to pay for the preferred present. Let the child know that you are there to help with, but not solve the problem. If the child feels "stuck," offer possible solutions. Make sure that the child makes the choice.)
- **Model problem-solving techniques:** Intermittently make a point of sharing out loud with students your problem-solving techniques. Doing this is especially productive if you happen to make a poor choice and they see that adults, too, make mistakes and have to deal with poor choices. You can do this "after the fact." (You had a flat tire on the way to work and you didn't have your cellphone with you.) Walk yourself through the steps aloud.
- **Teach kids to be on the lookout for barriers to problem-solving,** including

  not being willing to accept new ideas
  being too stubborn
  being too afraid of making mistakes
  allowing emotions to take control (getting too angry, upset, frustrated)
  seeing problems only as negatives, forgetting to look for their "good" sides
  blaming others and not accepting responsibility
  feeling sorry for self, whining, complaining
  manipulating and trying to wheedle out of solving the problem

One way to do this is to present to the class a problem, such as a friend telling you a lie; then, lead a discussion about the various barriers listed above. Invite students to talk about how each barrier could interfere with successful solving of the problem. In the case of "allowing emotions to take control," you would be unable to deal rationally with the situation, could become accusatory, and could be unable to accept or forgive.

- **Do take your role seriously:** You can help young people become lifelong problem-solvers so do your best to guide them. A good way to begin any unit on problem-solving is to share ideas about how people solve problems differently, and encourage introspection as to how the kids usually problem-solve. Having the skills necessary to problem-solve, at any age, in any situation, and to accept the consequences of those solutions, be they good or not-so-good, is what makes an individual confident, independent, and successful.

What to Avoid

- **Doing it all:** Don't jump in and make every decision for the child. It's okay for him to struggle, to make mistakes, even make bad choices. This is learning.
- **Being impatient:** Wait. Allow the child time to try. Sometimes adults, in their zeal to be helpful, prevent a child from working out the right or best solution independently.
- **Expressing disappointment:** Don't show displeasure with a child's choices, even though you may disagree with them. Unless the choice will have negative physical repercussions (e.g., choosing to eat only chocolate bars or to climb on a piece of school equipment not designed for climbing), step back and allow her to see for herself what her choice has wrought.
- **Saying "yes-but":** This attitude encourages self-doubt in the child, who naturally assumes his choice wasn't good enough.
- **Attaching blame:** Don't attach blame for a situation on anyone — not on the person experiencing the problem and not on any other person who might be involved. Deal only with the possible solution.

Seven Approaches to Problem-Solving

There are many ways to approach a problem. Although all of us use most of the approaches throughout our lives, depending on the particular problem, we veer toward one or two favorite approaches. Or we may use one approach first, discover that it doesn't work, and move to an alternative one. Each approach, or style, has specific tactics associated with it.

For the purpose of sharing approaches with children, I have selected and discussed seven frequently used approaches to problem-solving. In the last two cases, rational problem-solving is involved; in the first five, it may or may not be present, and if used, it's used erratically or poorly.

It is worthwhile to teach children about the various approaches, and encourage them to make appropriate use of all of them, even though they will eventually come to rely on the approaches they most prefer. By encouraging them to practice all the approaches, we will be giving them tools with which to more easily problem-solve as adults. They will quickly see the merits of some approaches and the weak points of others, and will be able to choose the one that best works for a situation, or, as mentioned earlier, to try more than one.

Helping young people to understand how they approach a problem and to identify what strategies they normally use (or don't use) is the first step in fostering improved problem-solving. Invite students to describe how they "tackle" a problem by first citing a specific problem (e.g., a broken promise). Encourage honesty. Self-disclose if you feel comfortable, perhaps adopting the Ostrich or a Cuckoo Bird approach outlined below.

Follow this introduction with a lesson or discussion about the seven listed approaches to problem-solving, all of which relate to the behavior of birds. Kids love birds — after all, birds can fly, and their quick, curious behaviors are fascinating to watch. Kids are familiar with the images of the wise owl, the early bird and the worm, the speedy hummingbird, and the singing nightingale. Capitalizing on this natural interest, you can introduce approaches according to the bird they are most like. Doing so usually captivates students and aids them in remembering all seven approaches.

Be sure to discuss how people often use more than one approach for a single problem. They might, at first, try to get someone else to deal with the problem, blame it on someone else, and then, maybe later, use a technique they have seen others apply to a similar problem.

1. The Ostrich approach

The ostrich hides its head in the sand and assumes it can't be seen. The primary goal of those who use this approach is to pretend the problem doesn't exist, to "hide the head in the sand" and hope the problem will go away. Occasionally, it does, and this intermittent reinforcement is often enough to encourage continued Ostrich behavior.

Pros:

- You don't have to deal with the problem right then.
- You can temporarily "escape."

Cons:

- Problem is still there — you can't run away from it.
- It won't go away on its own and may even get worse the longer it's left.
- Others may see you as a "chicken" or weakling.
- You probably won't feel good about yourself.

Example: Jack owes his friend some money that he borrowed for a cola. The friend expects to be repaid the following day at school. Jack avoids the friend all day.

What he needs to learn: You can't run away from problems. When you pretend they don't exist, they usually get worse.

What he could have done: He could have told his friend that he hadn't forgotten about repaying, but admitted that he couldn't pay yet.

2. The Canada Goose approach

Some Canadian geese wait too long to depart on their southward migration and get stuck in the ice and perish. The primary goal of those who use this approach is not to have a goal — yet; they want to delay having to deal with the situation. They convince themselves that "it can wait." Teachers may want to locate the fairy tale "Clever Alice" which describes a girl who spends so much time getting ready to do something she gets nothing done at all, as this is an excellent example of Canada Goose behavior. Often, procrastinating leads to more problems to solve. Procrastinators often have to face the Domino effect of the original problem.

Pros:

- Sometimes, waiting will help you to figure out a better way.
- You don't have to deal with the problem immediately.

Cons:

- The problem won't go away and if left may get worse.
- If you procrastinate in one area, you may be tempted to do it in other areas.
- Procrastinators are seldom as successful as people who take immediate action.
- Problems should be dealt with immediately (taking time for reflection can be part of this).

Example: Same as for the Ostrich approach above — Jack feigns illness so as to avoid facing his friend that day. As a result, he misses an important test that he'll have to write during basketball practice. Missing basketball practice will mean that he can't play in the next game.

What he needs to learn: Face problems right away. Putting them off allows them to escalate and doesn't make dealing with them easier. It can even cause more problems.

What he could have done: He could have thought ahead, weighed the options, asked for advice from his coach, and made a better choice.

3. The Cuckoo Bird approach

The cuckoo bird puts its eggs in another bird's nest and leaves the caring of the eggs to the surrogate mother. The primary goal of those who use this approach is to get someone else to take care of the problem. They immediately cry "help" instead of facing the situation on their own.

Pros:

- You are not alone; someone "shares" the problem with you.
- There is immediate relief from stress.
- The "helper" may be more knowledgeable than you.

Cons:

- Little self-confidence is developed — you likely feel helpless.
- Feelings of failure and weakness may accompany this approach.
- There is no growth in problem-solving abilities, no independence.
- Others get annoyed with your lack of problem-solving abilities.

Example: Consider the example of John, where he called his mother for help after he broke the expensive tool.

What he needs to learn: It's important to make decisions on your own. There won't always be someone to turn to; John needs to work on developing independence and on taking responsibility for his own actions.

What he could have done: He could have approached his employer with sincerity and offered to make restitution for the tool.

4. The Peacock approach

I will never forget the Grade 5 boy who came to school one morning with the following statement. "My mom lost her car keys and man, did she pull a Peacock!" All the other students knew exactly what he meant.

The peacock is notable for its showy display, which covers up a lack of any more aggressive moves. This approach, often called the "Drama Queen" approach, involves a dramatic, over-enactment of emotions accompanying the problem situation. The person panics, loudly and visibly, creating a visible and audible display.

Pros:

• Lots of attention, which may or may not be helpful, is forthcoming.
• Others immediately know of your distress; some may offer assistance just to defuse the situation.

Cons:

• People tend to reject this sort of extreme behavior.
• Little self-confidence is developed.
• The person is showing poor coping skills that may carry over to other situations, as well.

A good children's story that speaks to this problem-solving approach is "Chicken Little."

Example: The adolescent, Tara, panics when she loses her cellphone. Her loud display attracts her mother.

What she needs to learn: Panicking serves no one. It is an uncontrolled reaction that makes the situation seem even worse because the person — in this case, Tara — allows herself to *feel* and *act* out of control.

What she could have done: She could have adopted the "neutral mind strategy," which would allow her to think more calmly. (See page 17.)

5. The Robin approach

The robin is noted for its quick return in spring and its overzealousness in trying to find worms long before any are available. In addition, the robin starts to sing at false dawn, that is, before the real dawn begins. The primary goal of people who use this approach is to get off to a quick start; they jump in without thought, often with disastrous, or at least less than successful, consequences. When faced with a problem, the robins just want to "get it done." They seldom consider alternatives; instead, they act on the first idea that comes to mind.

Pros:

• The sooner you start, the sooner it's over.
• The problem isn't "hanging over your head."

Cons:

• Jumping in without thinking can cause serious errors in judgment.
• To problem-solve rationally, you need to take enough time to work through the steps.
• People may see you as being impatient and impulsive.
• Too quick a start usually leads to more problems.

Example: Richard rides his bike to school. On the way he falls and bends the front tire. He realizes he shouldn't ride it that way. He is already late for school and doesn't want a detention, though, so he rides the bike anyway, completely ruining the wheel and necessitating costly repairs. He has to pay for the repairs on his own since he made the situation worse.

What he needs to learn: Thinking through a situation thoroughly before taking action may be better than taking a "quick fix."

What he could have done: It would likely have been better to walk to school and be late rather than to do something that led to worse consequences.

6. The Peregrine Falcon approach

The peregrine falcon, once almost extinct, is an adaptive bird. It has managed to survive and adjust to a wide variety of "homes," including city buildings. It can live almost anywhere in the world, with the exception of the polar regions. People who take this approach to fixing a problem rely on a way that has worked before or has worked for similar problems. They problem-solve by referring to what is already known; they "adapt" familiar methods to the current situation. Fact finders, they use tried and true ways. They are methodical.

Pros:

- You can use a method that you know works.
- You don't have to take a chance or a risk.
- You will probably be able to problem-solve more quickly than by using the innovative method.

Cons:

- Since not every problem is the same, you may not have the exact solution for the specific problem, and hence, it may not work well.
- Overdependence on what you already know can lead to weaker problem-solving, especially if the situation is unique or unfamiliar.
- There may not be a reliable method in your memory bank, so total dependence on this style can leave you without a solution.

Example: Sal broke her pencil, again, by pressing too hard. She knew the teacher would be angry because she'd already sharpened it three times. Just like she always did, she decided to bother Peter into lending her a pencil and hope the teacher didn't see her.

What she needs to learn: If you are going to make use of a previously used strategy, it's important to ensure that it fits the situation.

What she could have done: Once before when she had no pencil, she used a colored marker, then rewrote the information in pencil later. This choice would have been better.

7. The Chickadee approach

The "resident" chickadee of Canada relies on innovative feeding behaviors and shows behavioral flexibility as a response to seasonal environments. The primary goal of people who favor this approach to problem-solving is to find an innovative solution to a problem; they tend to find a new or different way to solve the problem. They push limits and "think outside the box" to promote change. They are the greatest risk takers.

Pros:

- You are challenging yourself and thinking creatively — that's good.
- This style encourages the most interesting and unusual solutions and can lead to new, ground-breaking ideas.
- This exciting approach may yield surprising solutions.

Example: Instead of adopting the Peregrine Falcon approach, as described above, Sal picked the wood from around the pencil tip with her nail and her scissors; she didn't bother Peter or make her teacher angry, or have to rewrite the information later because it had been written in colored marker.

What she needs to learn: Sometimes, the best answer is the most creative (or crazy) one.

Tips for teachers on introducing approaches

1. A great way to begin a lesson about approaching problems is to have children jot down (or draw) words or illustrations explaining how they approach a problem when first faced with it. This begins with the teacher modeling the procedure by self-disclosing a real or imaginary problem, using the following steps.

    Problem: Lost some data (e.g., class attendance) on the computer.

    First approach: Ignored issue, hoping data would magically reappear later (Ostrich and Canada Goose)

    Second approach: Got really angry at the "stupid computer" and at own lack of technological savvy. Yelled and shouted a lot (Peacock)

    Third approach: Tried resetting the program, which had worked in the past, but failed to produce results this time (Peregrine Falcon)

    Once students have determined their own most used approaches, have them share, and add the bird names for the approaches where appropriate. For example: "I think you might be a bit of a Peacock." By using the birds here, you'll be creating interest in the lessons to come where you discuss each "bird" approach in detail. At this point, all the kids will know is that you, as the teacher, have identified a "bird" in what they've written or drawn. By doing this before the lesson(s) on the styles, kids will gain some personal insights. For example, a child may become aware that she often approaches a problem like a robin.

2. Provide students in groups of three or four with a problem, and have them create imaginary solutions, using each of the seven approaches. Have them discuss in their groups the pros and cons of each. Bring the class back together and collate ideas. One example of a good problem for group and class discussion is the issue of the class needing about $100 by a specific date for an upcoming field trip.

3. Open a discussion with the question "Is there ever a good time to use the Ostrich approach? the Canada Goose approach?" Encourage open-ended thinking and accept all responses, realizing that what is "real" for each child will differ.

    Possible answers:

    - Use the Ostrich approach when you find the problem overwhelming to deal with alone (e.g., child abuse); be an Ostrich until you can get help.
    - Use the Canada Goose approach when there is no possible way for you to deal with the problem at that particular time. (For example, you can't

deal with a broken promise at night when you are going out with parents; instead, deal with it the next day.)

Once children have an idea about various approaches to problem-solving, they are ready to learn the specific steps and skills of rational problem-solving so that they can incorporate these into their personal approaches.

Steps for Approaching a Problem

For purposes of instruction, this book offers rational problem-solving strategies in the following order:

1. Creating a Neutral Mind (or, Seeing a Problem Positively)
2. Evaluating a Problem's Seriousness
3. Using Rational Problem-Solving Strategies

Each of these steps is huge and needs to be dealt with separately. Once students understand them, however, they can use them quickly, instantaneously, and almost instinctively when faced with a new problem.

1. Creating a Neutral Mind (or, Seeing a Problem Positively)

It's easy to see a problem as a negative. It is, after all, a "bump in the road" associated with a negative emotion that if not dealt with, may eventually become a mountain. It gets in the way of the desirable easy life.

For the serious problems — the crises, the life-threatening situations — no positive ramifications can likely be discerned, at least at the onset. However, it may be possible to see positives at a later date. (I deal with serious problems in Chapter 6.)

However, whenever possible, it helps to think of a problem in a more positive manner. Problems make life interesting, exciting, stimulating; they help define a person's character. We all know that a little stress is good for us. Humans cannot live without some degree of stress. Stress is what "gets us going," what kindles our creativity, what arouses us, and what engages us in life. So let us suggest to our students that when it comes to problems they should readjust their thinking: that they try to see them as "positive stresses," as "little pluses," or as "potentially interesting bumps in the road." Let's encourage them to look at these hitches as blessings in disguise, at least whenever possible, and by so doing, they can likely deal with them more effectively. The list on the next page, presented as a line master, may help you to promote this positive thinking in your students.

Part of seeing problems in a new light — viewing them more positively — is in being able to understand the emotional climate of a problem-solve situation.

If someone is able to view a problem with optimism, desirable effects frequently occur. Teachers would agree that people who have a more cheerful outlook on life are better able to problem-solve. Mood affects how people process information and hence, solve their problems. Put more simply, people in a happy mood tend to be more creative and productive — that is, better able to problem-solve. The irony is that when people face a problem, their outlook tends to fade from happiness to frustration or anger, which impedes their problem-solving. Is it possible to avoid this development and thereby create a more constructive backdrop for accomplishment? I believe it is.

What does this mean for us as teachers in the classroom? It means that we can serve as models. When faced with a problem, we can model if not a "happy"

Seeing Problems Positively

- Problems can be the beginning of something new and exciting.
  Example: The problem of global warming has led to the designing of more efficient vehicles.

- Problems can help us to grow and develop our thinking skills.
  Example: The problem of coming up with a topic for a science report led to increased information about fragile ecosystems and how the student could help.

- Problems can be motivational.
  Example: The problem of wanting to be less timid around others prompted a student to ask to orally present the class's weekly news report.

- Problems can help us to meet new people.
  Example: The problem of moving and leaving friends led to meeting new friends.

- Problems can be opportunities for personal growth.
  Example: The problem of whether or not to attend church with parents led to an interest in learning about other religions.

- Problems can serve as beneficial learning experiences.
  Example: The problem of spilling blueberry juice on the dining-room carpet led to an Internet search for a stain removal.

- Problems can promote the creation of new ideas.
  Example: The problem of trying to make a baby brother stop crying led to the idea that maybe babies would enjoy being sung to in gibberish.

- Problems can make us stop, think, and take notice of our surroundings.
  Example: The problem of getting lost on the bike trail through the woods led to finding a new trail home.

mood, then at least a "neutral" mood. We can talk to our students about this, about the effect of frame of mind on the ability to problem-solve. We can share and practice with them specific actions that may reduce the negative humor that usually accompanies a problem (see the suggestions below). And we can consistently draw to their attention situations where another person's mental state has either positively or negatively affected an outcome.

The following are suggestions for achieving or maintaining the "neutral" mind.

- Take a deep breath with eyes closed. Hold for five seconds, exhale for five seconds, and do not inhale five seconds. Repeat three times.
- Wiggle your toes in your shoes while you silently count to 20.
- Make "mouth music" — any sound you can make with your mouth, such as quietly smacking lips, put-putting through pursed lips, or clicking the tongue — for 20 seconds.
- Quietly repeat any combination of nonsense consonant-vowel combinations, such as "la-la-la," "bah, bah, bah," or "ma, ma, ma," for 20 seconds.
- Squeeze hands into tight fists, then open, squeeze, and open again; repeat 20 times.
- Silently count backwards from 20 while keeping your eyes focused on a spot near the ceiling.
- Silently say a tongue twister, such as "A skunk sat on a stump and thunk the stump stunk, but the stump thunk the skunk stunk."
- Silently sing the lyrics to a favorite song.

Once a neutral mind has been established, the child has a better chance of correctly evaluating and dealing with the problem effectively and rationally.

Tips for teachers on creating a neutral mind

1. Invite students to brainstorm problems that ultimately turned out to be positive — something "good" came of the initial problem.
2. Share with students the list of possible benefits of problems. Have them work in pairs or in small groups to think of an appropriate example for each benefit.
3. Invite students to recall times when either they or someone they know faced a problem with an inappropriate frame of mind — maybe shouting, crying, or hiding — and what the result was. Encourage them to make up a name if they wish to protect someone's identity. Discussing what happens when a negative mood takes over can be enlightening — and often humorous.
4. Practice the different neutral mind suggestions with kids, one per week. Ask them to identify one or two favorites that they think will work for them; then, when problems arise, remind them to use their ideas for creating a neutral mind.

2. Evaluating a Problem's Seriousness

We face hundreds of problems every day. Most of them we deal with almost unconsciously — what to wear, who to speak to, what to eat, and so on. Then there are the minor disturbances, such as a late bus or an argument with a friend. These are short-lived annoyances, but the negative emotions attached to them make them problems nevertheless. There are also the more difficult situations that take some degree of focus to solve or at least to attend to, such as the loss of an object or rejection by a peer. Finally, there are the serious, potentially life-threatening problems that may have far-reaching repercussions, such as physical abuse or natural disaster. In all cases, the negative emotions that accompany the "problem" serve as motivation for the person to take action. Our job is to help children take safe and helpful action.

First, students need to understand that a problem is a negatively charged situation or predicament that requires attention. Once they realize this, they can be coached to categorize problems into degrees of seriousness. I suggest three levels:

Level 1	Level 2	Level 3
Mild (irritating)	Serious, but not life threatening	Life-threatening

Often, students have difficulty deciding just how serious a problem is. Although most can readily identify a life-threatening problem, they need practice in figuring out whether a problem is serious or mild. As adults, we need to reassure them repeatedly that all problems should be dealt with, but that not all will be dealt with in the same way or with the same expediency. The old adage "Don't sweat the small stuff" comes to mind and could be effectively shared with kids who tend to exaggerate the gravity of situations.

The following definitions may be useful.

Mild, Irritating: This type of problem can occur many times a day. Dealing effectively with such a problem makes life easier, but not dealing with it will not usually cause physical harm.
Examples:

- Forgot homework, notes, lunch, or locker number
- Lost bike key, lunch, mitts, or pencil
- Broke pen, bike, or piece of sports equipment

Serious: This type of problem might cause some injury, but it would be minor. The problem could escalate into a more serious one if not dealt with. It should be dealt with quickly, and usually or sometimes (depending on the age of the student) with the help or support of the teacher or other significant adult.
Examples:

- Lost wallet, identification, bus pass, cellphone, bike, or house key
- Took part in an illegal or hurtful activity (e.g., vandalism, shoplifting, bullying)
- Forgot phone number, specific responsibility such as babysitting, or appointment

Life-Threatening: This problem presents a threat of personal injury, even death. It must be dealt with immediately and usually requires adult assistance as soon as possible.

Keep in mind that what is a mild problem for one person may be a serious problem for another. For example, while most kids would find losing a lunch a mild problem, a diabetic child would find it a potentially serious problem. Group discussions will help children to understand the inconsistencies here.

Examples:

- Swallowed poison or something potentially poisonous (e.g., "bad" meat)
- Suffered personal injury (e.g., cuts, burns, falls, being bullied)
- Got lost in unfamiliar surroundings at night

Tip for teachers on evaluating a problem's seriousness

Invite students to brainstorm problems, real or imaginary; then, in small groups, have them list these under the specific headings according to severity. Follow with discussion and encourage justification of positions. Not all students will agree on where specific problems should be placed. Reassure them that the placing is up to each of them. Everyone is allowed an independent opinion.

Appendix A provides a table of possible problems that kids may face. This list, presented alphabetically, can be used here for categorization.

3. Using Rational Problem-Solving Strategies

In this book I recommend the use of five distinct steps for problem-solving. Each step should be taught separately; eventually, they will be interconnected to form the basic problem-solving framework. I am not suggesting that the steps in rational problem-solving be used for every little problem every day — that would be both time consuming and ridiculous. What I *am* suggesting is that teachers and significant adults intermittently teach, review, and utilize the steps so that they become ingrained in children's minds. In this way, children will retain them and be able to recall them as necessary throughout their lives. And isn't that what teaching is all about — helping the kids to no longer need us?

Rational problem-solving is the logical, common-sense, and sequenced way of dealing with a problem situation. It becomes a powerful process of learning, especially when people learn from mistakes. It is important to remember, however, that some problems have more than one solution, some problems cause more problems, and some problems cannot be solved.

For purposes of this book, the Five-Step Problem-Solving Plan will be as follows:

1. State the problem clearly (see Chapter 2).
2. Determine the possible choices (see Chapter 3).
3. Assess choices and select the best one (see Chapter 4).
4. Devise a plan of action (see Chapter 5).
5. Evaluate the success of the plan (see Chapter 5).

Since I believe that teaching and reinforcing each step individually is the best way to help children become proficient users of rational problem-solving techniques, the steps are presented one at a time, together with teaching tips and practice lessons. Each step, with the exception of the last two, takes a chapter.

Rational problem-solving, then, is designed to assist children in becoming independent, confident when faced with a difficult situation, and, of course, able to deal well with their own problems. We present the steps, practice them, reinforce them, point them out in all areas of the curriculum, and play games to further strengthen awareness of them.

Rational Problem-Solving and the Curriculum

Game Recommendation

A good game for beginning the unit or series of lessons on problem-solving is Pulling Pockets in which the pockets contain simple, Level 1 problems that call for simple, quick solutions. In this case, each child pulls a pocket and verbally shares a solution with the class. (See Chapter 10.)

Although I appreciate how busy teachers are and how impossible it is to fit everything into the curriculum, I recommend teaching a mini-unit on problem-solving, such as the one provided in Appendix C. Doing so will introduce the idea that "yes, you can problem-solve, and here are the easy steps to follow." The entire concept of rational problem-solving can then be easily incorporated into all areas of the curriculum.

Teachers wanting to add awareness about problem-solving will easily spot them, but here for ease of reference are a few ideas.

Health and Wellness: This is the most obvious place to include the mini-unit on problem-solving as the focus of this part of the curriculum is on living skills.

Social Studies: All the history-based units (e.g., pioneers, revolutions) offer many places for a quick discussion about how the various people used, or might have used, problem-solving techniques in order to reach their destinations, meet goals, deal with newcomers, and so on. The politics-focused units similarly offer myriad problem-solving situations. Simply ask students open-ended questions, such as "How do you think ___ solved the problem of ___? What steps might __ have taken? What other solutions might ___ have thought of before selecting ___?" In some places, a complete unit on problem-solving is a part of the Grade 3 Social Studies curriculum.

Science: This core subject naturally lends itself to problem-solving in the laboratory and experimental situations, but it also lends itself, for example, in animal studies. For example, you might ask, "What problem-solving skills help this animal to survive?" or "Pollution of the Great Lakes is a huge problem — how can we use rational problem-solving to come up with possible ways to improve this situation?"

Mathematics: Since this subject includes mathematical problem-solving within its content, I will not deal with it here.

Fine Arts: There are so many ways to specifically incorporate problem-solving techniques in these areas simply by posing the problems in such a way that students automatically use the steps. For example, you could say, "Use what you know about problem-solving to figure out how to best design the poster so that it does what you want it to." As another example, you could say: "In this scene the actor has to cry. This is a problem. Use rational problem-solving to find a way to do that."

Language Arts: In every story the students read, characters solve problems. Invite them occasionally to identify the problems and suggest the steps the characters might have taken to come to the solutions they did. Characters often make bad choices, which can make stories interesting. These are perfect times to draw to the students' attention how the characters might have chosen differently and to reword the problems together using the correct steps. Similarly, when writing, students can use the problem-solving steps to narrow down topics or ideas, or to find the best ways to begin or end compositions.

Physical Education: A natural place to incorporate problem-solving in physical education often arises during conflicts between students. Ask the students to use the steps to solve these problems without your help. Other related problems include requiring money for uniforms or bus trips; figuring out how to do homework and be at practice too; and improving skills to make a team.

2 Defining the Problem: Step 1

Six-year-old Emma approached her teacher with a frown on her face, and indicated she had a "very bad problem." The teacher naturally requested more information and the child said — I quote from the teacher's amused recollection — "Well, the gerbil died before dinner and I had to clean the cage after dinner, but Joey — that's my little brother — was crying so he didn't eat and he hit me, and I hit him back so my mom got real mad at me and told me not to clean the dead gerbil's cage because I hit Joey so now I got to do it, and there's this field trip tomorrow so I got a problem." The teacher was laughing when she explained to me that after much questioning, she finally discovered that the "real" problem had nothing to do with the gerbil, with Joey, or with the cage, but with the fact that the girl had forgotten her field trip money. She obviously needed help with problem defining.

This chapter is all about "getting to know the problem." Kids need to understand their problems, to see them from different perspectives whenever possible, and to be able to talk or write about them while they are still unsolved. They need to decide, first, if a situation really *is* a problem, how severe that problem is, and how to concisely, positively, and objectively state it either orally or in writing, and include within the statement, a goal for change. Often, this goal is inherent and not specifically stated, but children need to be reminded it must be there. Setting the goal is the hardest part of the whole problem-solve procedure. For example, instead of thinking, *My brother and I keep fighting all the time and Mom said if we don't stop, we can't go to the show on Saturday*, write: *In what ways can I stop fighting every day with my brother?* (The goal is to stop fighting.) The best problem statement is both specific *(stop fighting)* and measurable *(ways ... every day)*.

Instead of thinking, *I am failing math and I need to raise my mark this year*, write: *In what ways can I raise my math mark to 80 percent by June 1?* References to ways and June 1 are specific; the 80 percent goal is measurable.

Questions to Ask about a Problem

When teaching children about problems and how to correctly describe them, a good place to start is to have them answer some personal questions. The following questions help kids focus on what the problem is — and isn't: a good first step to correctly stating the real problem. Of course, using this questionnaire with every problem would take too much time, but using it occasionally will help develop effective lifelong problem-solving techniques. It is also presented as a line master at the end of the chapter.

Twelve general questions

1. Am I sure it really *is* a problem? Is it real or imaginary?
2. How big a problem is it? What level?
3. Can I solve this problem alone or do I need help?
4. What would happen if I just left it alone and didn't worry about it right now? In other words, how urgent is the problem?
5. Is there a time limit for solving the problem? Can I solve it quickly or will it take a long time, maybe several days or weeks?
6. Why is it a problem? Do I know?
7. What caused this problem? Was it something I did or didn't do, or was it outside my control?
8. Is this problem mine alone or do I share it with someone else?
9. Can I fix this problem completely? If I solve it, will it stay solved or will it recur?
10. Does this problem affect other people besides me? If so, who?
11. Do I have enough information to solve the problem?
12. What would my perfect solution be?

It's true that some children and adults alike see almost *everything* as a problem. We all know people like this; they make mountains out of mole hills, overanalyze, worry ad infinitum — and drive everyone else crazy. These people often benefit the most from a mini-unit in rational problem-solving because it enables them to finally discern the real problems from the trivial annoyances, and the problems from the everyday choices.

Three problem-search questions

Since, as indicated, the questionnaire shared above is unsuitable for every problem, the abbreviated "Three Problem-Search Questions" can be substituted for quicker use. The way to begin examining a problem is to ask each of the three questions and take the corresponding step. These questions and related steps are as follows:

1. Is the problem real or imaginary?	*Step:* Make a decision.
2. Is it mainly mental, physical, or emotional?	*Step:* Identify the problem's stance.
3. How can you word the problem?	*Step:* Write or state the problem.

These basic steps lead the child into the problem in such a way that he or she will be able to grasp the situation realistically and identify a goal. If any ambiguity or confusion remains, prompt the student to return to the 12 questions and work through them.

1. Is the problem real or imaginary? Step: Make a decision.

The first step in defining the problem is to decide if it is a problem. Plan to lead a discussion about identifying problems. Some possible discussion openers are as follows:

- What is a problem?
- Is every choice a problem?
- If I want to go to the washroom, is that a problem?
- This morning I broke a nail. Is that a problem?
- Oh dear, I just broke my pencil. Is that a problem?
- Turn to a neighbor and describe a problem you had this morning.
     (If kids say they didn't have a problem, ask whether they had to decide what to eat or wear, and if choosing could be considered a problem.)

The point is to get kids thinking about problems in general. It may work for your students to write about a problem in a journal or in a letter to a pen pal, or to make up a problem by discussing possibilities with a partner. It doesn't matter what they do as long as they come to realize that people face hundreds of problems daily, but only a few require them to really figure out what to do.

2. Is it mainly mental, physical or emotional? Step: Identify the problem's stance.

Advise students that once they have determined something is a problem, they would be wise to figure out the problem's main stance — mental, physical, or emotional — as this will affect the ultimate goal. Solvers should ask themselves whether the result, solution, or goal has mostly mental, physical, or emotional characteristics. For example, if a student's marks are too low to allow him to take part in a specific extracurricular activity, this problem has primarily a mental (brain challenge) stance. Of course it could be argued that loss of extracurricular activity then makes the problem both emotional and physical, but the key is that the solution to the problem involves mental processes. Although most problems have more than one stance, if students learn to identify the most obvious, it will aid in determining the best solution.

It is a good idea to clarify for students that problems can and do take in more than one stance. When they understand that the stance a problem takes refers to the nature of the activity involved in solving it, not necessarily how they are *feeling* about the problem, they will be better equipped for successful problem-solving. By sharing this information with our students, by teaching them how to determine whether a problem is primarily one of mental, emotional, or physical action, we empower them and give them yet another tool for gaining independence.

- A mental or academic goal suggests a brain challenge, a stimulation of the thinking and learning processes. Mental goals are associated with problems relating to schoolwork, marks, studies, and knowledge. The problems usually have solutions that focus on ways to improve knowledge, understanding, and even wisdom.
- A physical goal suggests attention to the body, to health, or to the altering of body image and appearance. Physical goals are associated with problems that relate to weight, strength, endurance, and physical skills.
- An emotional goal suggests attention to emotional control and to dealing with others. The majority of children's problems seem to fit this category, and since strong emotions are usually associated with problems, it can be difficult to separate this type of goal from the other two types. An

emotional goal can include everything from learning how to deal with anger to learning how to say "no."

3. How can you word the problem? Step: Write or state the problem.

Albert Einstein once said that if he had one hour to save the world, he would spend 55 minutes defining the problem and only 5 minutes finding the solution. This idea reminds us of the importance of taking time to really understand a problem, to look at it from more than one perspective, before jumping in to try to solve it. Because of the importance of this step — and the fact that it is often the most difficult to follow — it will be considered at length.

The significance of correctly worded problems cannot be overestimated. Many kids have trouble putting what is "wrong" into succinct, specific words; this is where teaching comes in because solving a problem or at least coming up with workable ideas is always easier once the problem is clearly stated. Children who have clearly defined their problems will attack them more confidently.

Once kids know for sure they have a problem, they can apply strategies, which we can teach, to help them clearly define it. Of course, I am not referring here to those myriad little annoyances that plague us daily — what I will call the "choices" — but to the more difficult, Level 2 and Level 3 problems that require specific strategies.

Problem-Defining Practices

In order to simplify the problem-solving process so that kids can easily grasp and hang on to it, I will use five familiar steps for problem definition. These steps are (1) stating, (2) chunking, (3) clarifying, (4) framing it positively, and (5) formulating a question. Following them just to define the problem may seem like a lot of work, and initially it is, but doing so is well worth the time and effort. Eventually this procedure becomes second nature to children and they can quickly move through the stages to come up with a perfectly defined problem. Remember what Einstein said about thoroughly understanding a problem before trying to solve it.

1. State the problem

Teacher Tip
Invite kids to find a "talk-to thing," any inanimate object to which they can tell their problems. A favorite toy, figurine, or pet rock will work. An interesting art idea for creating the "talk-to thing" is to stand light bulbs in a base of clay, then paint faces on them, add hair, ears ... any personal touches, and give them names. These "talk-to things" readily become silent listeners for problems. My students named them "worry willies."

Say out loud the problem as you see it. For example, a high school student says: "I am trying really hard to lose weight because I feel ugly and fat, and I don't look good in shorts and we have to wear shorts in gym. My friend is having a big party and there will be lots of junk food like soda and chips and stuff that will wreck my diet, but if I don't eat that, everyone will tease me and know I'm on a diet. But I really want to go to the party."

The act of stating aloud serves an important function: It helps to clarify and organize thoughts, making possible solutions easier to formulate. This idea is certainly worth sharing with kids. At this point, rambling is okay, even encouraged. The idea is to "get it all out" and then go from there. This initial definition of the problem is done as if explaining the situation to someone with no knowledge about it. Details are included and necessary.

2. Chunk key words together

2. Chunk key words together

Teacher Tip

Since "chunking" is a valuable skill for many areas of the curriculum, teach and reinforce it regularly. A good practice is to provide a paragraph and have students circle or highlight only key words.

Kids are generally familiar with the concept of "chunking," or taking out unnecessary words and putting together those that work together. During a teacher presentation accompanied by some form of text, they could look at your words, or if they have spoken aloud, they could rethink their words to "chunk" key words together. Doing so will help them concisely outline a problem. Although rambling is encouraged in the stating step, here it is discouraged. At this point, good sentence structure is irrelevant. Students search for the most important words and sometimes substitute a word that encompasses several of the originally used ones, for example, *diet, party, junk food, avoid embarrassment.*

3. Clarify and include goal for change

The wording now becomes important. The original words or thoughts are re-examined, clarified, and objectified. Since chunking has already occurred, students should have only key words that they now make into a plausible sentence defining the problem. By its very nature, this sentence should include a goal for change. *At the party I don't want junk food to ruin my diet but neither do I want to feel embarrassed.* (Goal: Attend party without being embarrassed)

4. Make it positive

Teacher Tip

A mini-lesson on being positive is always useful. You could share any children's story in which the protagonist is a positive person, despite obvious difficulties. Invite children to suggest how the story would have been different had the protagonist not been so positive. An example is the familiar children's tale *The Little Engine That Could*, by Watty Piper. Another is the wonderful poetry book *If You Could Wear My Sneakers* by Sheree Fitch, illustrated by Darcia Labrosse, and produced in association with UNICEF. The poems in this book are based on the UN *Convention on the Rights of the Child*. Each poem represents someone with a problem, such as a sloth being differently abled with three claws; however, the animal has a positive outlook, and the poems end on a positive note.

The problem with problems is that, by their nature, they imply difficulties and carry negative auras. So, at this point students can rewrite or reword the problem so that it reads or sounds as positive as possible. Sometimes, making a problem sound positive seems impossible, but even the idea of looking more positively at a situation has merit. Consider this sentence: *I will feel better about myself and my diet even while attending the party.*

5. Turn the problem statement into a question

Turning the problem statement into a question will help kids move toward a solution. Our brains naturally want to answer questions. The question generally leads to the action plan. The following pointers will help kids to do this.

Follow the previous steps (state the problem, chunk, clarify, and make it positive) until you have the goal at that stage. For example: *I will feel better about myself and my diet even while attending the party.*

Brainstorm ways to turn this *declarative* sentence into a question: *Why will I feel better if …? How can my diet make me feel better …?* Most of the time, adopting the phrase "In what ways …" is a good start.

Examine the brainstormed ideas and pick the one that does the best job of covering all the key points from the earlier declarative sentence.

Add to or delete from that selected question until you feel it has fully covered the situation. Thus the declarative sentence could evolve into the question *In what ways might I deal with the party food so that I will feel better about myself?*

(The girl who had this problem solved it beautifully by bringing a veggie platter to the party, as well as a six-pack of diet soda.)

Even when intentions are good, it's not always possible to word a problem exactly according to these criteria. That's okay. As long as children are aware of the practices for correctly defining their problems, they will incorporate enough of the points to make their statements and questions objective and clear.

Teacher Tip

One of the most valuable skills we can develop in our students is the ability to ask or write good questions. To promote this, provide a number of declarative statements and invite students to turn them into questions. Here is an example.
Declarative sentence: I am unhappy because I have no friends in this new school.
Question: In what ways can I make new friends so that I will feel happier?

Game Recommendation

A good game to reinforce specific definition techniques is Name Game. (See Chapter 10.)

Teacher Tip

As an excellent follow-up to learning how to write problems well, invite kids to either write e-mails to imaginary people telling about their imaginary problems, or to write to "Dear Abby," the imaginary person who will answer problems with possible solutions. Have students begin by writing the detailed, rambling problem (Step 1) and end by writing the succinct, clear, positively framed problem question. Example:
Step 1: Dear Abby. I have this terrible problem that is driving me crazy and keeping me awake all night every night. It's my brother. He keeps getting into my room and taking my stuff and my stuff is private but when I tell mom she just tells me to keep my stuff put away and to clean up my room but he still goes in there. I have no privacy and I want to kill my brother sometimes. We are always mad at each other and I hate that. What can I do?

If, after going through the steps to better defining a problem, some children still struggle, you may offer them possible ways to start defining a problem positively. Make a wall chart with prompts such as those that follow or provide a line master so that children have an instant visual reminder. Once they get the idea, your students will be able to add to this list.

A Word about Wording

The way a problem is worded, with thought for the connotative nature of chosen words, is a final consideration for Step 1 of the Five-Step Problem-Solving Plan. I have not mentioned wording until now because depending on their ages and abilities, it may be confusing for some kids and hence not useful to their teachers. However, if you decide that your students can understand connotative wording, consider adding it to your lessons on defining a problem.

Some words generate negative feelings. A friend who is a nurse commented to me on how all "procedures" in the operating room used to be called "operations." The latter word is no longer used because of the "bad feelings" it gave patients. The same applies to the word "problem" — it has a negative connotation. A problem, by definition, is *a difficult situation, a hindrance, a crisis, a drawback.* None of these words sounds positive. Consequently, looking for alternatives to the word "problem" is a good idea.

The following alternatives to the word "problem" work well with children. I am not suggesting that any one of them will work consistently for all problems, but an awareness of them, and perhaps a concentration on one or two, really helps to see the "quandary" more constructively.

- *Speed Bump:* A speed bump is a nuisance, but it serves a positive purpose — it helps us to slow down. A Level 1 or Level 2 problem may be viewed as a speed bump, something that encourages the involved person to slow down and make a wise choice.
- *Situation:* The word "situation" seems to have a calming effect. A situation is neither bad nor good; it is neutral and so instantly removes the pessimistic feeling from the problem.
- *Opportunity:* Rewording a problem to call it an opportunity (for advancement, for improvement …) makes it appear beneficial and removes the inherent difficulty that goes with a problem.
- *Challenge:* Kids usually love challenges, as long as they are not overwhelming, so calling a problem a "challenge" can motivate them into action more readily.
- *XXX:* The term *XXX* is a space holder for a word of the child's own choosing. With a Grade 5 class who seemed to have more than their share of "problems," I found it helpful for the kids to choose any word they wanted, nonsense or otherwise (as long as it wasn't profanity or someone's name), and dub all future problems as such. One young man chose "Mugwort" (inspired by the Harry Potter series, I believe). The next day he calmly told me how he had solved a tough Mugwort the previous night and I understood instantly. Some of the words chosen were *toadell, zakey, ferret, didlyacat, jushmush.* Use of such words helped defuse some of the negative emotions usually associated with confronting a problem.

Possible Problem-Definition "Starts"

Some students struggle to get started writing. Others may find it hard to get down to defining a problem. When it comes to problems, here are some "starts" that you may find helpful.

- In what ways might I … (better than "How can I …")

- What might I change to …

- How might I create …

- In what ways might I combine …

- In what ways might I fix …

- What options might I have to …

- In what ways might I adapt or change what I already know to …

- What might I take away or remove so that …

- In what ways might I proceed …

- In what ways can I overcome …

- In what ways can I say …

- In what ways can I make a difference to …

- In what ways can I collect (provide, share, give …)

- What might I do to make (myself, my dad …) feel (stronger, better, healthier, happier …)

- _____

- _____

- _____

Twelve General Questions to Ask about a Problem

1. Am I sure it really IS a problem? Is it real or imaginary?

2. How big a problem is it? What level?

3. Can I solve this problem alone or do I need help?

4. What would happen if I just left it alone and didn't worry about it right now? In other words, how urgent is the problem?

5. Is there a time limit for solving the problem? Can I solve it quickly or will it take a long time, maybe several days or weeks?

6. Why is it a problem? Do I know?

7. What caused this problem? Was it something I did or didn't do, or was it outside my control?

8. Is this problem mine alone or do I share it with someone else?

9. Can I fix this problem completely? If I solve it, will it stay solved or will it recur?

10. Does this problem affect other people besides me? If so, who?

11. Do I have enough information to solve the problem?

12. What would my perfect solution be?

3 Determining Possibilities and Choices: Step 2

Grade 3 students were trying to solve a class problem. In groups, they were brainstorming for ways to deal with a couple of kids acting up and teasing other kids as soon as the teacher left the room for a few minutes.

To help the students come up with many possible solutions, the teacher invited them to "be creative," to include even ridiculous, impossible solutions on their lists — and they took her suggestion to heart. When groups were asked to share their lists, one group, amid giggles, offered the idea that as soon as the teacher left the room, everyone (except the "teasers") should pull out a filled water gun and start shooting at the would-be teasers. The class had a good laugh and one student added that the water-shooting strategy "works to keep my cat from climbing on the table — we squirt her just a little with water and she gets right down."

In this step in the Five-Step Plan, students seek as many possible solutions to the problem, in as many possible ways as feasible. It is the generation-of-options stage, the free-flow-of-thought stage where no suggestions are considered incorrect, silly, or unacceptable. Every idea is considered. Practicing it will help students overcome the tendency to depend on their first idea.

An important point to remember here is the huge effect of mood on the effective free flow of thought. If the potential problem-solver is in a "bad" mood — the usual result of facing a difficult problem — creative thinking or any kind of productive thinking may be impaired. If dealing with a child who is upset by or angry with a problem, first encourage the child to establish a positive or at least a neutral mood (see page 17). If none of the neutral mood options work, recommend that the child go out and play or do some other physical activity, such as walking, before trying to think creatively about the problem.

The other impeder to creativity and productive thought is fear, which is why people in life-threatening situations often attempt unsuccessful solutions or grasp at the first solution that comes to mind. The best advice to share with kids here is to take a few deep breaths — if there's time — before jumping at a solution. (I will deal more with life-threatening problems in Chapter 6.)

Once the prospective problem-solver has dealt with these factors, it's time for the free flow of ideas and open-ended searches for a wide variety of possible solutions. Ways to stimulate ideas and find possibilities include these:

- brainstorming or brain-writing
- inviting ideas
- asking a senior
- surfing the Net

Beyond encouraging adaptive thinking, or recall of "what has worked in similar situations," our job as teachers is to encourage divergent thinking and to reinforce always that many options should be considered.

Games for Brainstorming

Point Please? Students look at familiar situations from unusual viewpoints. The teacher, for example, provides the idea of a rainy day. Students, in groups, brainstorm alternative ways to think about a rainy day. Examples: Sad for picnickers, happy for ducks, good for gardeners, bad news if river is high, bad news for roofers, sad for worms. *Excuses, Excuses!* Pairs of students brainstorm excuses for common situations, striving to come up with more ideas than other pairs. Take, for example, the situation of undone homework: Dog ate it, Mother laundered it, little brother got sick over it.

How to Brainstorm

1. Accept any and all responses.
2. Avoid discussion at this time.
3. Keep a record of responses.
4. Set a time limit and watch the time.
5. Don't overlook the obvious response.
6. Take turns; listen to everyone.
7. Build on another's response.
8. Focus on quantity, not quality.
9. Don't judge — there are no wrong answers.
10. Keep an open mind and have fun.

Teacher Tip

To best incorporate both adaptive and innovative thought, recommend that kids either divide a page in two with a vertical line, or use two different pages. On one (side), they can jot the out-of-the-box innovative ideas, while on the other (side), they can list ideas that they know have worked. This combination of brainstorming and brain-writing allows kids to see that good solutions can be derived from either or both.

Brainstorming or Brain-Writing

In brainstorming, individuals or groups orally toss ideas around; in brain-writing, these random thoughts are jotted down in any shorthand understood by the child. Since most problems belong to one person specifically, brain-writing is often the best approach. If the child has become familiar with the idea of talking out loud to a "worry willie" (see page 26), he or she can certainly brainstorm to it. In either case, if the child is trying to solve a problem independently, there will be limited verbal interaction at this stage.

To help children understand and utilize this step in problem-solving, we must teach the how-to steps within it and also model them. Keep in mind that during this step, there are no right or wrong responses. During this productive time, quantity matters more than quality and creativity is encouraged.

If a child tends to adopt the adaptive, or Peregrine Falcon, approach to problem-solving, she will probably come up with ideas that are familiar, that she's witnessed being effective in similar situations. Accept these while at the same time encouraging thinking "outside the box," being "wild and creative" even if the thoughts seem too fantastical to be useful.

Similarly, if a student tends to take the innovative, or Chickadee, approach to problem-solving, it's important to encourage the child to think of any similar situations where an already proven solution might work.

Everyone is familiar with brainstorming; however, I have found that a few simple steps make the experience more effective. These are outlined below:

1. The teacher "sets" the experience by saying or writing and then rereading to the class the situation. Example: "The problem we will be examining is this. We want to help our neighboring school earn money for a new playground. They need a lot of landscaping done, as well as the purchase of new equipment. They aren't as lucky as we are. Our playground is fantastic, wouldn't you agree? How can we help them?"
2. The teacher reminds or invites students to say aloud the rules of brainstorming. (See "How to Brainstorm" in the margin.)
3. The class is given silent think time in which they focus on the problem at hand. Closing eyes is helpful here. Usually 60 seconds is enough.
4. The teacher breaks the students into groups of four or five and provides the start cue and a time limit for the brainstorming. Setting a time limit encourages students to think more quickly. Talking out loud within groups is encouraged.
5. When less than one minute is left in the set period, the teacher lets students know.

Brainstorming stimuli

We have all had the unfortunate experience of "blank-slate-head." We actively seek possibilities but come up with nothing. Kids experience this, too, and we need to reassure them that this empty-head feeling is perfectly normal. (I once had a child who confided she was sure there was something terribly wrong with her brain because she had been brainstorming "forever" and hadn't arrived at a single thought.) Taking quiet time and playing brain games are ways of overcoming a mental block.

A wonderful Chinese proverb speaks to this QT position:
Inner silence promotes clarity of mind;
It trains us to go inside
To the source of peace and inspiration
When we are faced with problems and challenges.

Game Recommendation
Students can also play the game Brain Blast, which promotes divergent thinking. (See Chapter 10.)

• Taking Quiet Time

A suggestion that works for me — and for most people, including children — is to enforce a quiet time by going for a walk alone (the preferred tactic), sitting in a quiet, secluded spot, or if at home, lying in bed (if you can do so without falling asleep). During this time, the person with the problem keeps thinking about it, but doesn't worry about coming up with possible solutions. You just keep bringing the mind back to the problem, as you've defined it. Amazingly, the mind will, as a rule, start popping ideas to the fore. Often, the first idea leads to a second and more. With kids, I refer to this procedure as "taking QT." Many of them can even do this at their desks, for example, when working on a tough math problem or coming up with ideas for a creative writing project or journal entry. I find thinking in a very "alone" space most effective.

• Playing Brain Games

Sometimes, brainstorming can be unproductive. Perhaps, the problem situation is either incorrectly defined or is out of the students' experiential realm. For example, a problem relating to future employment would not likely bring many responses. Sometimes, however, the dearth of suggestions may be the result of what my kids dubbed "boggled brain." By this they mean that their thought processes are "stuck" and need saving and waking up. So, if the teacher feels that the problem is well delineated and students *should* be able to provide more possibilities than they are, there are a couple of games that may boost creativity: Action Alphabet and Crazy Talk.

Action Alphabet: In groups students try to think of any possibility that begins with the next-in-sequence letter of the alphabet.
PROBLEM: *In what ways can I overcome my fear of talking in front of the class?*
Student 1: **A:** "**A**fraid, I feel afraid so I should take a big breath."
Student 2: **B:** "**B**ack wall. I hate looking at eyes so I could look at the back wall.
Student 3: **C:** "**C**alm ... I need to look calm so I will stand straight and pull my shoulders back."
Students continue in this way until the end of the alphabet. If a student cannot come up with a word to match a letter, the student may say, "pass." The student with letter "X" could choose a word that begins with any letter. Action Alphabet is a fun and stimulating way to boost creativity.

Crazy Talk: In groups of five or six, students make up crazy words with crazy meanings that might be possible solutions. For the same problem noted in Action Alphabet, for example, students came up with these words:

- "Bemansing," defined as "Be like Mr. Manse," who was their favorite teacher and who was always well prepared
- "Talk-sock," defined as "talk their socks off"
- "Jimmy-C-be," defined as "talk using your mouth like Jim Carrey"
- "Spaily," defined as "speak slowly like a snail moves"

A lot of fun and laughter are involved with this approach, which lets kids see that sometimes wild and wonderful responses have purpose. The facts behind these silly, crazy talk responses are valid.

Teacher Tip
Put kids in groups of five or six, and give each group one sheet of paper. The first person jots down a possible solution, then passes the paper to the next student, who reads what was written then jots down a different possibility, and so on. Students continue passing and jotting until no one can think of anything different. At that point, they discuss the possibilities together. With this method, students' thoughts are stimulated by what they read from peers.

Inviting Ideas

Sometimes, kids feel they are alone with their problems, so we as adults can remind them that seeking assistance can be a wise and mature thing to do. They just have to remember that the problem and the solution are their own. So, when a child has brainstormed and done some brain-writing, and still doesn't have a viable solution — or perhaps just wants more input before choosing a solution — encourage the child to ask peers, parents, and trustworthy adults for input. Drawing on the wisdom of others cannot be overlooked.

We can teach kids *how* to ask for help. In other words, they should learn not to say, "Can you solve my problem?" or "Tell me what to do." A question or statement like either of these speaks of dependency. Instead, we can teach them ways to phrase their requests, for example, to say: "I'm working on a problem. What ideas can you share with me …?

Let students work in pairs to practice appropriate ways to ask for help; ask them to make a note of their favorite ways. Then, have the whole class share ideas so that students can create repertoires of approaches.

It is a good idea to caution students against asking for help as a first approach, though. Remember that we are teaching independence, *not* co-dependence.

Asking a Senior

Sometimes, we forget how much seniors know, how many years they've had to absorb ideas, and how many great ideas they have to share. When teaching kids about ways to gather information to aid in solving problems, be sure to remind them about this valuable resource.

Asking a senior for advice is different from inviting ideas from parents, teachers, or peers. Seniors often have different suggestions. For example, say the problem involves needing money. While a peer might say, "Ask for an advance on your allowance," a senior might say, "Perhaps you could help the gardener here at the home and he'd pay you a few dollars." Seniors also tend to have more time to listen and sometimes their calm listening serves to help the child generate more ideas. Finally, seeking out a senior frequently involves the child making a commitment to go somewhere, such as the senior's home or lodge. This, too, may prompt idea generation. It certainly speaks to the child's dedication to the task. Most kids have either an elderly neighbor or a grandparent at least as close as the phone, if they are unable to make physical contact.

One lucky 11-year-old, Becky, lived near a seniors' lodge and shared this story with me.

Becky had a "situation." She needed money to buy an iPod to take to camp with her and her parents had promised to share the cost 50/50. She didn't want to save all her allowance because she used it on weekends with friends at the mall. She defined her situation in this manner:

In what ways can I make $55 by June 30?

After brainstorming and brain-writing, and asking her friends and parents for input, she still felt she lacked an appropriate solution so she visited the seniors' lodge. They knew Becky — she was a frequent volunteer. One senior told her the following: "Well, when I was your age, I darned socks and did laundry for extras, but I guess people don't darn socks any more." Becky was excited. No, they didn't darn socks, but her mom did do laundry and hated ironing so Becky would do the ironing for a fee. Becky hugged the granny and went off with a happy solution to her problem. As it turned out, Becky's mom was thrilled with the solution, too.

When teaching kids to ask a senior, there are a few things worth mentioning. Here are some of them:

- Be sure to clearly state your problem, and be prepared to provide the additional details the senior(s) will probably want. (Remember that many seniors are hard of hearing.)
- Take paper and pencil with you (or have them beside the phone), and jot down any ideas that may prove valuable.
- Ask questions if you don't understand a response.
- Remember to be polite and to thank the senior for the help, even if the suggestions were not helpful for your particular situation.
- If you did use a suggestion, be sure to let the senior know how it worked out. Everyone loves to hear about a success, especially if they have contributed to it.

Surfing the Net

I don't need to say much about this except that I discourage it as a first attempt at finding a solution. Remember that we are building lifelong problem-solvers who can face and successfully manage problems daily, and even though technology is moving at light speed, not everyone will have instant access to the Internet. With the wealth of information available there, though, it would be ludicrous to say, "Don't use it."

The biggest concern is that using the Internet can often take more time than budgeted — it's easy to get carried away. With most problems, the solutions are required quickly. So, if the student can let the problem wait and needs more possibilities, then recommend use of the Internet. An example of this type of problem is where a child has two weeks before the visit of an aunt to come up with a way to tell her that he has lost the expensive watch she gave him. The student can type in the keywords from the problem definition. With the case of the lost watch, he could type in words such as "apology," "apologizing," "accepting consequences," or "accepting responsibility."

To help students avoid getting caught up in surfing the Net, advise them to limit their time to 30 minutes. If they have not come up with possible solutions to a specific problem during that time, then students should leave the computer, try other approaches to generate ideas, and then, if necessary, return to the Internet for another 30-minute search. By limiting the time — set an alarm or timer — a student is less likely to get sidetracked. Books, such as *Internet for Kids: A Beginner's Guide to Surfing the Net* by Ted Pederson, may provide helpful advice, as well.

Consolidating Steps 1 and 2

At this stage of problem-solving, a good idea is to pose a class problem and follow through the steps together. Once you've defined the problem according to the steps outlined in Chapter 2, then move to Step 2, determining possibilities. Draw attention to and practice the various approaches, but do not evaluate any possible solutions yet. Simply gather and record them.

Below is a sample scenario.

- State the problem as the need for money to hire a bus to take the class on a field trip, while not wanting to ask parents for more money since they've just paid for something else, like special supplies.
- Invite students, in small groups, to reword the problem properly and to ensure that a goal is included. For example, In what ways, without asking parents, can we earn money for the bus for the field trip?
- Brainstorm for possible solutions. Assign different groups different approaches. Groups 1 and 2 might brainstorm and brain-write. Groups 3 and 4 might invite information and ask advice from seniors — for example, with prior warning, arrange to send them to the principal, to a nearby seniors' home, or to another teacher or librarian. Group 5 might surf the Internet.
- Compare all thoughts and ideas as a class. For example, students might report that seniors suggested bike riding or finding parents to drive; students surfing the Net might have got the idea of asking for a special rate from the bus company, perhaps as a goodwill measure. Collate the ideas as much as possible, but do not evaluate them yet.

4 Narrowing and Making Choices: Step 3

In a Grade 5 class, the students were brainstorming possible solutions to a class problem. The teacher noted that one child had removed himself from his group and was sitting at his desk, head down between his arms. Thinking he must be ill, she approached the boy and asked him what the matter was.

"Oh, I'm sick all right," he replied. "We've come up with so many possible solutions that my head is bursting. As far as I'm concerned, we've just made an even bigger problem because now we'll never make the right choice. I'm taking a time out on this one."

In another incident shared with me by a Junior High teacher, two students at Step 3 of problem-solving had a conversation much as the following.

Student A: Okay, too many choices so I choose not to choose.

Student B: No way. You can't do that. That's a cop-out.

Student A: Yes, I can. Not to choose is a choice, too. I can be in the middle.

Student B: Well, my dad said that people who stand in the middle of the road get run over …

Choosing can be difficult, but we can teach our children tactics to help make the best choices. First, a good rule of thumb is to come up with about 10 possibilities on which to make a decision. At this step in problem-solving, children scan and assess the amassed possibilities. Doing so is much like looking at a plate of freshly baked chocolate chip cookies and being allowed to take only one. The cookie connoisseur may carry out a thorough scan to assess cookie size, number of chocolate chips per cookie, and degree of "doneness." This analogy works well with people of all ages.

Children, like adults, need time to re-examine, reflect, and assess each possible solution. This step should not be hurried. Sometimes, "sleeping on it" helps; children can collect possibilities, then choose not to choose at that time. They can choose to make their choices on a specific day, or at a specific time — of course, doing this is possible only if the problem doesn't require immediate action. With most "big" problems, taking the time to make the best choice is usually preferable to jumping in and possibly implementing a solution that may cause further problems.

Children as problem-solvers can commit to memory a few basic steps to help them critically examine their stock of ideas so that they come to one "best" solution. These steps are (1) limiting choices, (2) prioritizing choices, and (3) making the final choice.

1. Limiting Choices

If kids succeeded in the previous step of divergent thinking, they may now face so many possibilities that the task once again seems impossible. We need to reassure them that *too many* is better than *not enough*, and that having quantity allows them to make a better final choice. For this step where the number of ideas must somehow be reduced, here are some suggestions to help with that task. This information is presented in line-master format for students at the end of the chapter.

Ideas to pass on to students

1. Ask yourself, "How much time do I have to solve this problem? Can I think about it for a while? For how long?"
2. Scan all ideas with an open mind, crossing off any that are too obscure, obviously unworkable, or silly.
3. Now, look at the remaining possibilities and ask questions such as the following of each. Not all of these questions may need to be asked in every situation — they serve as a guideline only.
   • Can I do this? Alone? With help?
   • What do I need to implement this suggestion?
   • Is this possibility anything like I have tried before? Why did or didn't it work?
   • What are the possible positive and negative consequences of this choice?
   • Will this choice solve the problem?
   • Is my problem "goal" mainly physical, mental, or emotional — and does the solution I've chosen fit this goal? (Sample problem: In what ways can a student improve his math mark to 80 percent? Solution fit: The goal is mental, so the chosen solution must reflect mental development.)
   • What's the worst that could happen if I chose this possibility and it backfired? Could I live with that?
   • Can I make this choice alone or do I need some adult (professional) guidance? (Be sure to let kids know that they can't solve all problems and that it's okay to seek help with a tough decision.)
4. If you like the solution, ask yourself what might make it hard for you to implement. For example, if it requires money, as for tutoring, do you have it or can you get it? If it requires time, such as you would need to learn a new skill, will you have enough?
5. Are there any other solutions that you also like that could be combined with this one to make it even better? In other words, can you use more than one suggestion, or can you slightly change the "best" choice to incorporate others as well?

   No. 1 choice: Get a tutor.
   Other choices: Work with a peer, deliver papers to earn money to help pay a tutor, set regular weekly study times.
Can you implement these possibilities, too?

2. Prioritizing Choices

Once the possible solutions have been limited to no more than three, it's time to prioritize, or determine which option is preferable based on knowledge of the problem, the situation, and personal capabilities. The ultimate goal of ranking possibilities is to understand which possibility is most likely to provide 100 percent success.

Consider this problem based on Grade 4 social studies: A class in an otherwise fairly heterogeneous community wants to celebrate the arrival of a few new students from different countries, while ensuring that what they do will be positive for everyone. After discussion, they limit possible solutions to (1) creating multicultural posters, (2) inviting the parents of the new kids to come to class and talk about their home countries, and (3) hosting an open house, where all the parents and students can meet. Based on their responses to the questions, the class decides on possibility 3. Earlier in the year, they had held a successful open house for parents so they knew they could make all the arrangements alone, with just the teacher's help. They saw their goal as physical (getting parents to come to the class, arranging chairs, etc.) and also mental (thinking up activities, making introductions, etc.). They thought they could act on possibility 1, as well, by decorating the room with multicultural posters. For many situations, one solution appears much more feasible than others.

But what happens when kids can't decide or when more than one "probably good" solution is available? In these cases, it often comes down to a virtual coin toss as to which option is tried first. The good news here is that should solution number one fail, there will still be numbers two and three to try.

Alternatively, what happens if, after detailed consideration, students decide that none of the possible solutions measures up? At this point, they should ask for advice, take a break, or return to a neutral mind, and begin brainstorming again.

Here are two tidbits to share with your young problem-solvers.

1. When you can't decide — and you've worked through all the steps — ask for advice.
2. Try writing out in detail each possibility. This perceptual motor activity is often enough to shed light on the solution.

3. Accepting Responsibility

To make a choice is to accept responsibility, to take control. Some children, though, do not yet understand what personal responsibility and control mean, and have not internalized them. As a result their confidence in their own choices is rather shaky. They make a choice and then right away doubt it. Immediately, they think that they have made the wrong choice, that any other choice would have been better. Second-guessing is one of the most unproductive human behaviors.

Once again, the teacher or another concerned adult should step in and provide reassurance that all choices are good choices when made with an awareness of the facts and with honest approaches. English author George Eliot said, "The strongest principle of growth lies in human choice." Kids need reassurance that they can make good choices. They also need reassurance that if their choices aren't so good, they have the strength, ability, and responsibility to face the consequences.

Here, in brief, is what teachers and other significant adults should tell kids about facing and accepting their choices.

1. If you have "owned" the problem, then whatever choice of solution you make will be okay. A sense of calm follows any decision, even a poor one, so appreciate it.
2. Determination to follow through with your number one choice is important, even if that means "taking a risk." Life is full of risks. We don't move ahead, grow, and evolve without them.
3. Trust your "gut feeling," otherwise known as "intuitive decision making." (Often kids are leery about adopting this method of decision making, so teachers should encourage them to do so, especially given that they have already established priorities.)
4. Avoid second-guessing. Make a decision and stick with it. Constantly worrying that you might have made the wrong decision causes further anxiety and wastes time. Trust in yourself. If you have made a mistake, you can deal with the consequences because you foresaw the possibility and are prepared; or, if you have made a mistake, you can correct it because you are a problem-solver.
5. Consider how this decision will affect others. Who are these others? (Often, the act of identifying other people is enough for egocentric kids to see the possible effect(s) of a decision.)
6. Everyone makes mistakes. It's often how we learn. An unknown author wrote, "Good decisions come from experience, and experience comes from bad decisions." So you see, either way — you win!

Making that big decision as to which solution to implement can be carried out fairly quickly once students are familiar with the process. Again, not all the ideas suggested in this chapter need be used. Just present children with enough choices that the ones they come to like, the ones that work for them, will stand out and stick with them long after they have left our care.

How to Go about Limiting Possible Solution Choices

Once you have identified a problem and a range of possible solutions for it, you are ready to evaluate your options and determine the top few choices. Here are some ideas and questions to consider to aid in this process.

1. Ask yourself, "How much time do I have to solve this problem? Am I able to think about it for a while? For how long?"
2. Scan all ideas with an open mind, crossing off any that are too obscure, obviously unworkable, or just plain silly.
3. Now, look at the remaining possibilities and ask questions such as the following of each. Not all of these questions need to be asked in every situation — they serve as a guideline only.
   - Can I do this? Alone? With help?

   - What do I need to implement this suggestion?

   - Is this possibility anything like something I have tried before? Why did or didn't it work?

   - What are the possible positive and negative consequences of this choice?

   - Will this choice solve the problem?

   - Is my problem "goal" mainly physical, mental, or emotional — and does the solution I've chosen fit this goal?

     Sample problem: In what ways can a student improve his math mark to 80 percent?
     Solution fit: The goal is mental, so the chosen solution must reflect mental development.

   - What's the worst that could happen if I chose this possibility and it backfired? Could I live with that?

   - Can I make this choice alone or do I need some adult guidance?
     Know that it's okay to seek help with a tough decision.

Preferred solution: _____

4. If you like the solution, now ask, "What, if any factors might make it hard for me to adopt this choice?" For example, if it requires money, as for tutoring, do you have it or can you get it? If it requires time, such as is needed to learn a new skill, will you have enough?
5. Are there any other solutions that you also like that could be combined with this one to make it even better? In other words, can you use more than one suggestion, or can you slightly change the "best" choice to incorporate others, as well? Example:
   No. 1 choice: Get a tutor.
   Other choices: Work with a peer; deliver papers to earn money to help pay a tutor; set regular weekly study times.

Other possible solutions: _____

Can you implement these possibilities, too?

5 Acting and Evaluating: Steps 4 and 5

Jason was a great problem-solver — on paper — but in reality, he may well be one of the world's best procrastinators. When I got to know about his no-action skills, he was in Grade 8; however, the students in the class had been aware of his tendencies for years. When it came to generating ideas, possible solutions, probable causes, and so on, nobody could top Jason. His mind whirled like a top and he spouted out responses more quickly than I could jot them down. However, when it came to selecting one and putting it into action, Jason quit. He always said, "I'm a thinker, not a doer." He certainly was that.

Then came the day Jason had a problem of his own. His mom had expected him to babysit his younger brother, but he had soccer practice and if he missed it, he wouldn't be able to play in the playoffs. With Jason's permission, as a class we used the problem-solving plan and generated many possible solutions to his problem. We narrowed it to two good solutions and left it up to Jason to make the final choice and put one into effect. When he didn't show up for practice, we assumed he had chosen to babysit. But when his mom phoned the school to find out why he wasn't home to babysit, we realized Jason had "done it again." Instead of acting on a solution of his choice, he'd avoided both possibilities — he had played video games with a peer.

For most people, adults and children alike, formulating a plan of action is less difficult than fulfilling the preceding steps. By now the problem is understood and the best possibilities for a solution have been outlined. All that remains is to devise and implement a plan that, it is hoped, will lead to the desired solution.

It is important to delineate the specific steps that will be taken to reach the goal, to break the solution down into manageable, easily evaluated components. For example, if the initial problem related to improving marks at school and the chosen solution was to increase study time to one half-hour daily, the student could follow this up with a targeted plan as to how, where, and exactly when that half-hour daily study will occur. There should also be a timeline so that the student will know when to re-evaluate the situation. For example:

After week 4, May 12, monthly tests should show a raise in marks of 10 percent.

Students may also need to be told that not all problems get solved overnight; many take years.

Implementation Strategies for Students

When instructing our students about the components of problem-solving, we as teachers can share with them a number of strategies beyond outlining steps and timelines. We can also invite students to choose one or more before putting a plan into action. These strategies will help children broaden their problem-solving bases and enable them to make independent choices in the future.

- Jot down exactly how you plan to proceed — this is like making a shopping list. This perceptual motor activity encourages remembering so that you can proceed more easily.
- Follow this with a more accurate "good copy" of your action plan, and keep it in a safe place for future evaluation. This step is very important in the problem-solving process.
- Draw charts, illustrations, tables, and graphs to represent the solution. Doing this is like drawing a map of how you will get to a particular spot.
- Close your eyes and imagine yourself following your plan to the end. Try to foresee possible pitfalls and consider how you will deal with them. This envisioning is like a "dress rehearsal."
- Work backwards. Think of what you want the end result to be, then move back from that one step at a time to recheck each preceding step. Doing this is like unknotting a rope.
- Recall a similar problem you've encountered and solved, and try to remember the steps. Will they work for this problem? If so, then you've already had a practice run. If you had trouble solving the previous problem, then you have an idea of what to avoid trying.
- Write an open-ended sentence about how you'll feel when the plan is completed. For example: "When I … [insert steps in plan] I will feel …" This exercise in positive thinking may give you the impetus needed to fulfill your plan.
- Use a calendar to write on and mark successes. Clearly note the date on which you expect to successfully meet your goal and thereby solve your initial problem.

The first point in the above list — jot down how you plan to proceed — is the only one that kids have to carry out, especially when they are just learning about problem-solving. Eventually they will do that step only, or mostly, in their minds, but initially, putting thoughts on paper is important. For those kids who, like Jason, are perpetual procrastinators, the visual outline of an action plan is often enough to jolt them into activity.

Understanding an Action Plan

Sometimes, devising an action plan is difficult. Sometimes, kids even ignore this step, leaving the results of their previous steps "hanging" and their self-confidence in jeopardy. At this point kids already have a clear goal in mind, have considered many alternatives for reaching that goal, and have challenged each viable alternative to arrive at a final few.

One way to help kids get a better idea of an action plan is to compare it to a road map. The starting point is where they are at the beginning of their problem solution; the end point is where they want to get to. They plan the entire "route," but are alert to possible detours or roadblocks — things, situations, people — that may interfere with the smooth movement toward their end points. This analogy seems useful especially when a problem is big and has a lengthy timeline.

An action plan typically includes exactly what the child plans to do, when, and how he plans to do it. The design of the plan depends on the initial problem and final goal. The following points help clarify what an action plan is and what it does.

Action plans

- specify the actions (strategies) needed to reach a goal
- identify the specific, measurable objective of each strategy (if desired; not always appropriate, especially for younger kids)
- include a specific timeline
- state an exact starting date, time, and, if appropriate, place
- specify input from others (e.g., adult supervisor, peer), if necessary
- are "written" or recorded in an easy-to-read format

The following is an example of an action plan created by a Grade 6 student.

Goal	Strategy	Objective	Responsibility	Timeline
Increase math mark to 80%	1. Study ½ hr. nightly, in kitchen. 7–7:30 p.m.	1. This will help me catch up & remember formula. Chapter test scores up 5% by end April	1. Mom will remind me. I'll set stove timer.	1. Will do for one month. Start April 1
	2. Attend teacher's tutorials every Mon. 3:30	2. She will help me understand & review. Fri. test scores up 2–5% each week	2. Mark on calendar each time I go. Record test scores.	2. Until end of tutorials (end May)

Developing an Action Plan

There are a number of contexts in which a teacher should step in and provide the impetus for children to move ahead with an action plan. One is, quite naturally, at the appropriate time when teaching a complete unit on problem-solving. Another would be when a student or group of students has asked for help in implementing a solution for a specific problem. A third might be during a group discussion centred on actions taken by a story protagonist to help solve a problem. There are other situations where talking about action plans would be appropriate, but these tend to be the most common.

When teaching students how to develop an action plan, be sure to introduce the following steps. This information is presented as a student line master at the end of the chapter.

1. Write your goal clearly. (What do you want to accomplish?)
   - A stranger should be able to read the goal and know exactly what you mean.
2. List each of the up to three possible solutions you have selected from your list separately, specifically, and concisely.
3. Determine which kind of plan you will need. Note that some action plans require sequenced steps, where one part must precede another. For example, if learning to skate, you would put "can skate forwards without falling" before "skating backwards." Other plans, as in the above study chart, require several "actions" to be carried out simultaneously.
4. Make sure that each "part" is clear and that you can keep track of it.
   - Break the steps into small, easily reachable "bits." A weekly goal may be better than a monthly one (e.g., improving weekly math test scores by 2 percent), even when the ultimate goal is more distant (e.g., earning a mark of 80 percent in math).
5. Include a timeline.
   - You may find it helpful to use a calendar to keep a running record.
   - Include checkpoints, where you can evaluate how you are doing.
6. Evaluate your progress at regular intervals.
   - It's important to see how close you are to your goal by reviewing what you have already accomplished. Be accountable to yourself.
7. If necessary, modify the plan as you go.
   - If you have to make a change, do so; then, return to the plan. (A family activity might make attending a Monday tutorial impossible. A modification to the plan might be as simple as changing the night of the tutorial or as complex as coming up with a flexible tutorial schedule together with increased one-on-one time.)

Game Recommendation

The game Disorder helps reinforce the importance of correct order or sequence when carrying out almost anything, including an action plan. (See Chapter 10.)

It's important to help kids connect their plans right back to their original problems, and to reassure them that these action plans are fluid. They are guidelines only, easily changed if necessary. Remind kids that if they find themselves "sliding backwards" or "following the wrong path," it doesn't mean that their plan has failed. It is merely a momentary hindrance, a manageable setback. The most important thing is to get back on track as quickly as possible.

Evaluating Success

At this stage, the success, or degree of success, of the attempted solution will be evaluated. Since sometimes a solution fails, we need to teach for the possibility of either success or failure. How does one determine the accuracy of either and what can be done in the event of the latter? There are two parts to evaluation: one ongoing, the other summative. These will be dealt with separately, although both may well be a part of an action plan.

1. *Ongoing Assessment:* Kids are familiar with this; it happens all the time at school. But when do they use it as opposed to an "end result" evaluation? If the action plan calls for a period of a week or more, a good idea is to do some ongoing assessment. The following questions that students can ask themselves are helpful.
   - Am I moving toward my goal?

- If not, why? What do I need to change?
- Do I need to add or take away anything from my plan?
- Am I progressing, but at a slower pace than expected? If so, can I change this or do I have to accept it?

2. *Summative Evaluation:* This concept also is familiar to kids who have had to deal with pass/fail results all their school lives. We just need to remind them that every action plan worth carrying out is worth assessing at the end. Since they have gone to all the trouble of creating this solution to a problem, they should be glad to acknowledge their success or even their failure. (Point out enthusiastically how this is a learning situation.) Posing the following questions is helpful.
   - Did you get the result(s) you wanted?
   - Did anything unexpected happen? Was this good or bad?
   - If you didn't get the results you wanted, what can, or will, you do now?

Teacher responses for this last question could be as follows:

- "Don't worry. Don't panic."
- "Look carefully at your plan and try to figure out what went wrong."
- "Review your original brainstorming possibilities. Maybe you'll see a better choice you overlooked the first time."
- "If you are unsure about what to do, give yourself a full day, 24 hours, without thinking about the situation; then, after that, return to it."
- "Still confused? Make sure that the original problem *was* the real problem, and not the symptom or consequence of a different problem. At this stage, you may need to get help from an adult."

A point is necessary about the final teacher comment "… symptom or consequence of a different problem." For kids, this is a difficult construct to understand. What it means for them is that they are trying to solve the wrong situation, and consequently their results don't work. You don't want to head into the realm of psychology with your problem-solving kids. Teachers confronted by kids whose action plans have failed to achieve the desired outcomes may want to consider the following suggestions.

Ways to Help Students with "Failed Solution Situations"

- Provide reassurance. We all fail sometimes.
- When possible, use humor and self-disclosure to defuse the child's negativity if it is apparent. "Well, this didn't work. What a total nuisance, eh? I'll bet you're mad as a hatter. I know I'd be spitting fire about now!"
- Remind the child about "neutral mind" and positive statements about self.
- Try to determine the cause of the failure. Is it a lack of or error in information? How can you help with this? If it is a flaw in the plan, you can work through this with the student. For example, if the student's goal was to improve his mark by a specific date, and you note that his action plan was unrealistic and he has not given himself sufficient time to accomplish the goal, you can point this out and suggest a more realistic time frame.
- If the failure is due to an entirely different situation (e.g., a student's parents are divorcing and the student is trying to solve the problem of his mom's

anger toward him) and you can see that — even if you can't identify the situation — use your best professional judgment. Suggest that the child "leave" that particular problem and focus on a different one. Help the student choose a problem that you know can be solved more easily.

- Be aware that the failure to solve a problem might be a defence mechanism at work. Some children have difficulty with success. They seem to almost unconsciously set themselves up for failure, as if success would be painful or detrimental to them. There is little a teacher can do about such a situation other than be aware of it and seek professional guidance for the child. If you suspect this, you will need to discuss the matter privately with the student or with the school counsellor. As a rule, a situation in which a student seems consistently unable to successfully solve an identified problem probably has psychological implications. Professional intervention may be required.

Example: Allie, 10, was working on a problem that involved fighting with her siblings and knew that her mom wanted to stop getting involved in these petty fights, which she found upsetting. Allie had an excellent action plan worked out, but at the end of the one-month period she had designated, she tearfully admitted that she was fighting more than ever. The teacher was surprised until she did some homework and discovered that Allie's stepsister had recently moved in with them, and Allie was having a hard time accepting her mom's affection for the girl. By fighting with the stepsister, as well as with her other two siblings, Allie was probably acting out her inner turmoil; if she was able to stop fighting, there would be no further outlet for her frustration.

Action Plan Outline

1. Write your goal clearly. A stranger who reads it should know exactly what you mean and so should you.

   Goal: _____

2. Record up to three possible solutions that you have selected from your list. These should be your strongest ideas. List them separately, specifically, and concisely.

   Possible solution 1: _____

   Possible solution 2: _____

   Possible solution 3: _____

3. Determine which kind of plan you will need:
   - sequenced steps, where one part must precede another *or*
   - several related "actions" carried out simultaneously (e.g., tutorial as well as daily half-hour home study)

   _____

   _____

   _____

   _____

4. Make sure that each part of your plan is clear and that you can keep track of it. Break the steps into easily reachable "bits." For example, set weekly goals (e.g., improving weekly math test scores by 2 percent), even when the overall goal is more distant (e.g., earning a mark of 80 percent in math).

   _____

   _____

5. Include a timeline, perhaps making use of a calendar to keep a running record. Identify check-points at which you can evaluate how you are doing. Checkpoints are not pre-set; they will vary according to your plan and your goal.

   _____

   _____

6. Evaluate your progress at regular intervals.
7. If necessary, modify the plan as you go.

Modifications:

   _____

   _____

6 Facing Big, Scary, Life-Threatening Problems

A young boy, I'll call him Mike, shared this story. Mike, an 11-year-old, was playing in the dense bushes on his grandfather's farm, a fair distance from the house, when he came upon what he thought was a wolf, busily devouring "something dead." He froze. His heart thumping, he took a big breath then slowly backed away until he reached a tree, which he hurriedly climbed. Once he was high enough that the wolf, then snarling at the foot of the tree, couldn't reach him, he started screaming for help. His grandfather soon arrived and chased away the animal, which turned out to be a feral dog, probably more dangerous even than a wolf. The grandfather confirms the story and proudly explains how his grandson was "an amazing problem-solver."

Another true story of a youngster who displayed excellent problem-solving skills in a life-threatening situation is the following. This boy, Carl, was only nine when he heard his new baby sister crying during the night. He wondered why his mom wasn't going to see to the baby. Upon entering his parents' bedroom, Carl found them both asleep and he wasn't able to waken either one. He acted quickly; he opened the window in his parents' bedroom then ran to the baby's room and took her outside. Once outside he called 911 on the cellphone he'd grabbed on the way out. When asked later why he behaved in this way, he explained that he'd learned at school about the dangers of carbon monoxide poisoning and that was all he could think of in the situation. As it turned out, he was right and his act of opening the windows probably saved his parents' lives. The paramedics awarded Carl an "honorary membership" for his skill and bravery.

Molly, a 12-year-old, was babysitting her younger sister when she heard a strange noise in the house. The doors were all locked, but she was sure she'd heard the breaking of a window in the kitchen. She quickly and quietly ran upstairs to where her little sister was sleeping, grabbed the child and her cellphone, and locked herself and her sister in the bathroom. Once safely locked in, she called 911 on the cellphone. The police arrived to find that the house had, indeed, been broken into and that Molly's excellent problem-solving may well have saved her and her sister from injury. The would-be thief was apprehended and Molly became somewhat of a town hero.

A final story I would like to share is that of my husband, Jerry, as a mere five-year-old. He was walking on thin ice — literally — wearing a one-piece snowsuit, heavy boots, mitts, and so on when he fell through the ice. He remembers in detail the sudden fear that gripped him. He also recalls thinking that he had to stay calm and look for the break in the ice, which at that moment wasn't readily apparent. He felt himself sinking, but instead of panicking, he moved his arms slowly — hard to do given the weight of his sodden suit — and finally saw the break in the ice. He remembers carefully, so as not to break any more ice, dragging himself out of the frigid water and lying prone on the ice surface for "what seemed like hours." He then remembers thinking he'd freeze to death if he lay there, so he crawled and hobbled to the nearest house. To this day no one can quite understand how a five-year-old could have solved the problem in such exemplary fashion.

In all of these cases, the children displayed excellent problem-solving skills, even when faced with frightening circumstances. How is it that some kids can behave so calmly, so appropriately, while others panic, run, scream, and frequently make the situation worse? I don't have an answer, but I do have a few suggestions that will enable more children to problem-solve effectively in life-threatening situations.

What Constitutes a Really Serious Problem?

First, what is a Level 3, or life-threatening, problem? It is any situation that has the potential to cause bodily harm or mental harm, to either the child or someone close to the child. I am not a psychologist so cannot speak to the enormous emotional trauma that can result from one of these terrible problems; what I can do is provide some guidance for teachers or significant adults who want to help children solve problems as effectively as possible, no matter what their nature. The most common of these problems include

Teacher Tip

When discussing Level 3, or life-threatening, problems with kids, it is a good idea to ask them what kinds of problems they think fit this category. If they miss any of the ideas outlined in the text, then add them to the discussion. This discussion should be kept as positive as possible. Be on the lookout for any unusual reactions by students, as these may indicate a problem to deal with in more depth later.

- abuse, physical or mental
- attack by an animal or another person
- abandonment or being lost "anywhere" and alone
- natural disaster, such as being caught in a flood or a tornado
- severe reaction to potentially poisonous liquids, inhalants, or drugs
- severe injury, perhaps from a bike or auto accident, or a fall

Teachers and other significant adults are responsible for discussing with kids how to react if they ever find themselves facing any of these severe problems. The steps to successful problem-solving don't necessarily work here — time may be too tight. In these situations, the first thing is always to get help. However, what about most of the kids in the incidents outlined above?

- If Mike had called for help first, the animal would probably have attacked.
- If Carl had called for help first, his parents might have died.
- If Molly had dialed 911 first, she or her sister might have been hurt by the thief.
- If Jerry had struggled, he would probably have sunk more rapidly.

How did these four children manage? They all "assessed" the situation before they did anything else.

I recommend that all kids be taught the following ABCD approach to those scary life-threatening situations. They may not use all four steps or any for that matter. They may be able to go directly to step 3 — call for help — and if so, that's great. But having these steps firmly implanted in their heads and being able to call on "something" instantly can be helpful in emergencies. The ABCDs of solutions for emergency problems are easy to remember.

The ABCDs of Serious Solutions

A — **A**ssess the situation: Quickly do a visual scan to see exactly what's happening.

B — **B**e safe yourself: Always consider your own safety first.

C — **C**all for help: Shout, scream, make noise, use a cell or land phone.

D — **D**on't leave: Remain where you are, if possible.

These steps, because they follow the ABCD sequence, are easily recalled in emergencies, but need to be discussed and practiced with kids so that they become familiar with what each one encompasses.

Assess the situation

Teacher Tip

Share one or all of the previous incidents with kids and point out how each child — Mike, Carl, Molly, and Jerry —"assessed" the situation. Invite kids to think of other instances, real or imaginary, where "assessing" is important.

This idea, borrowed from First Aid and CPR courses, suggests that the child takes a good look before jumping in or reacting in any way. It is the take-a-big-breath step. In the incidents described at the beginning of the chapter, each child did this, and by doing so, avoided panic. Assessing is a rapid, instantaneous action, not a prolonged fact-gathering one. It basically gives a first impression of the situation, something that too many of us fail to do before reacting.

Be safe

Teacher Tip

Share any or all of the opening incidents with students and invite discussion. Ask if the children could have reacted in other ways that would still have ensured personal safety. With older students, you can use the idea of a person downed by electricity and how, if the area isn't safe, the would-be helper would be electrocuted by touching that person.

Once the child has done an assessment, the next thing is to ensure personal safely. Carl did this by climbing a tree. Mike did this by moving outside. Molly did this by locking her sister and herself in the bathroom. Jerry did this by remaining calm. In all four cases, the children did their best to ensure personal safety — or, in the cases of Carl and Molly family safety — before calling for help. Carl thought of his parents; Molly's interest was for her sister, not just for herself.

Kids' immediate thought may be to scream for help, but if they are taught to see the importance of safety first, it may save their lives. Once again, it is a quick safety check that comes before seeking assistance. What, for instance, might happen if the assistance doesn't arrive? If the child has missed the personal safety step, and no help comes, personal safety may be in jeopardy. Of course, there may be instances where this step is skipped. I'm not sure, for example, whether Jerry used this step consciously when trying to find the ice opening. Kids need to understand that not all the steps need to be used, but they should at least be considered.

Call for help

Teacher Tip

Brainstorm who and where to call for help to aid children in making personal lists of people to turn to for help. They may want to include the family doctor, veterinarian, police, poison control centre, next of kin, and reliable adults (other than kin). Prompt each child to create a personal phone list and then complete the list with accurate phone numbers. The portion of the list noting emergency numbers for kin and other adults should also be prioritized, with the first person the child would call being identified first.

This is the most natural step, and unfortunately, it is often the only step used by those with weak or no problem-solving skills; however, in the case of any Level 3, life-threatening problem, calling for help is mandatory. We all face big problems that we can't deal with alone. We all need help. We all need, at times, to call upon the skills and wisdom of others. In a serious situation, it's important that kids seek help.

Of course, in some situations, such as family abuse, kids may hesitate to do this. These are the cases where it is illegal for teachers or significant adults not to report their concerns to the authorities, such as social services, the police, or the medical experts — these are not the types of problems that can be dealt with in this book. However, kids can, and should, be encouraged to seek help as step 3 of the ABCDs of dealing with severe problems. In several cases, I have found that just being taught about the ABCDs has provided the impetus for abused children to take this important step.

Do not leave

Teacher Tip
Ask whether any of the students has ever been lost, and if so, what they did. Open a discussion about being lost alone in a forest with night coming and "no service" on the cellphone. This scenario allows for discussion about all four ABCD steps.

This step is important especially for some problem situations, such as being lost, trapped, injured, or frightened. The natural tendency to "flee" can be the downfall of some people in severe problem situations. If, for example, Molly had run out of the house, the thief may have injured her or her sister. If Mike had turned and run, the animal would probably have given chase and injured him. Although neither Molly nor Mike stayed exactly where they were, they showed discretion and considered safety first; they didn't leave the general problem scene.

This step becomes even more important when someone is lost. If, for example, a child has taken the first three steps, but is still lost in a park or forest, the best choice is to stay put. Children need to be aware of this step but also to understand that it may not always apply.

A little planning goes a long way when teaching children how to deal with scary or life-threatening problems. As teachers, we are faced with many situations that readily lend themselves to opening discussions of this nature. Examples include an upsetting news story with which students are familiar, the reading of a book where the protagonist experiences a scary situation, a real-life situation involving one of the students, and a social studies or historical passage. As usual, answering kids' questions as openly and sincerely as possible will help them to better build their own problem-solving repertoires. It's important to share with kids that some problems, for example, those related to health or weather, cannot be readily solved; some problems, including mental illness and abuse, need input from professionals; and some problems belong to a whole group of people and not just to the child (e.g., family problems, gang situations).

By teaching children tactics, we are giving them other ways to make sense of their environment and, perhaps, to control it. We know that children will meet more and more serious problems as they grow. We help them become problem-solvers so that they can be active participants in our expanding world and face even the big, scary, life-threatening problems with a measure of calm and independence.

Lean on Me: A game adapted from my earlier book, *3-Minute Motivators*, makes a good follow-up to the challenging and often upsetting lessons about serious problems. Partners "freeze" on cue and lean together such that one is always using the other for support — that is, "A" supports "B" or "B" supports "A." Partners alternate in providing support.

The teacher randomly calls out positions, each indicating a specific emotion which will qualify the way students freeze and lean together. For example, if the teacher calls, "depressed," partners will freeze with one supporting the other and both displaying depression with their faces and bodies. On a new teacher cue, such as "happy," "anxious," or "scared," partners move again and then refreeze, displaying the emotion. To improve the game's challenge, students can work in fours with two students leaning at a time.

This game requires debriefing. It's good for kids to express how they felt when being supported or providing support. The teacher can then link this to getting support for difficult problems.

Dealing with Children's Worries

Problems, worries, anxieties, and anger are all a part of life; however, the amount of each, the degree of anxiety, for instance, that each individual feels, differs greatly. We all know those people who never seem to get upset, who are forever grounded and calm no matter how terrible the situation. Similarly, we all know the worriers, the edgy types, the people who seem afraid of — or at least timid about — almost everything, and who often tend to overreact. Children show these tendencies, perhaps even to a greater degree, and they have fewer coping strategies than adults do.

To many of us, worries are seen as problems when, in fact, they are usually the results of problems, the symptoms of underlying problems. As teachers, we often witness these symptoms in children before we know what the problems are, so sometimes it is helpful to first deal with them.

Remember, though, without worries and problems we would fail to grow as sensitive, functional, viable human beings. Kids need to worry a bit, too. However, if the worries seem to verge on acute or overwhelming, then it's time to step in. Numerous sources and resources deal with childhood worry but for quick access, here are a few signs that may indicate excessive anxiety in children.

Symptoms of anxiety

- Restlessness, insomnia, sleep problems such as nightmares, sleepwalking, constant waking — manifested at school by overt demonstrations, such as constant yawning or head dropping to desk, and tiredness
- Severe self-consciousness
- Unexplained panic or phobia
- Obsessive compulsive behaviors, such as constant handwashing or non-stop pencil tapping
- Constant need for reassurance from a significant adult, even for simple tasks
- Unexplained bouts of crying, temper, anger, acting out
- Unexplained and unusual fears of people, situations, or objects
- New and unexplained nervous actions, such as nail-biting, teeth grinding, tics
- Daydreaming, lack of focus
- Unusual doodling, scribbling, or writing
- Destructive behaviors, such as destroying books or articles of clothing

This list is by no means complete, but will serve to give teachers a quick check.

If concerned about a child, it is important to enlist professional help. No one should attempt to deal with a situation he or she may not be qualified to handle, though. If unsure whether you can discreetly go about soliciting help for a child or whether parental permission is necessary, ask the school counsellor or principal. Protocols differ from school district to school district, but as a rule, strict guidelines are set and must be followed. If, for example, you suspect child abuse, you are required by law to take action, such as first talking to the principal or counsellor. On the other hand, if you suspect nervousness, talking to the parents may be the best first step.

There is no one way to deal with a child's worries. There is no one-size-fits-all or single cure or even "best" intervention. Similarly, it's possible to have a number of ways to solve a problem or cope with an anxiety. So, it is the adult's responsibility to do three things:

1. Remind or inform kids that problems, anger, and worries are normal and not necessarily bad or wrong.
2. Let them know that there can be several different ways to solve a problem or cope with a nasty situation. Whatever works for them is the right way.
3. Under no circumstances should the problem-solving or coping strategy involve violence. No act of self-abuse or abuse of others is ever accepted. This is 100 percent non-negotiable.

The single most useful thing an adult can do for an anxious child is to find a way to help the child release some of the stress, the worry, the anxiety. One highly recommended method of doing this is through physical activity, which can be as simple as stretching fingers, arms, or whole body, or as energetic as running around the schoolyard. Once the stress level has decreased, then try to identify the underlying problem, using questions such as those in "Questions to Ask about a Problem" in Chapter 2.

Anxiety relief ideas

Here are some recommended ideas for helping kids to stop worrying:

- playing a game of checkers, basketball, cards — as long as it's not alone
- using the fine arts: drawing, painting, coloring, playing an instrument (if able), dancing, listening to music, or reading a good book
- writing about worries: using a personal journal, writing to a real or imaginary friend
- volunteering to help younger kids, seniors, neighbors, and others
- talking to a trustworthy adult, possibly a teacher, counsellor, coach, or relative (Some children will follow this route easily; others may need coaxing and encouragement.)

To introduce the anxiety relief ideas, share a worry — it can be imaginary — and then ask kids what you might do to "take your mind off it." Many of the ideas may fit with the above list. Share the anxiety relief ideas, by reading them to the students or providing students with a handout, and together create a brief wall chart that focuses on the verbs. For example, the first point, "playing a game of checkers, basketball, cards — as long as it's not alone," would become "playing with others."

You could also have students list their worries. If they say they have none, ask them to make up one. (An invented worry is usually based on a real one.) Then, ask them to match an anxiety relief idea with each worry. If they are comfortable doing so, encourage sharing with a partner.

As adults, we know that worry is just another form of fear, an insidious, dangerous form that is truly a time thief. Time is precious; our students are precious. Let's help them get over, or at least lessen, their worries. We can initially help by being observant and noticing when a student seems anxious; then, we can intervene in some way to assist in lessening the anxiety and help the child find ways to handle and reduce anxiety on his or her own. Teaching mini-lessons about anxiety-reduction ideas, posting wall charts on possible ways to deal with anxiety, and sharing books such as *What to Do When You Worry Too Much: A Kid's Guide to Overcoming Anxiety* by Dawn Huebner and Bonnie Matthews or *David and the Worry Beast: Helping Children Cope with Anxiety* by Anne Marie Guanci and Caroline Attia, as well as having group discussions, are all valuable and effective teaching approaches.

As a final word about those big, scary, life-threatening problems, tell students over and over again that the key word is HELP. They should get help, ask for help, shout help, enlist help. These problems they can't deal with alone.

7 Building Students' Self-Confidence to Facilitate Problem-Solving

Julie was an intelligent girl who seldom did anything wrong. In fact, she was terrified of making a mistake or doing poorly. Once when she got an unusually low mark on a test, she burst into tears, refused to accept the paper back, and then refused to write the subsequent test. Julie had a problem, but when the teacher attempted to walk her through the problem-solving steps, Julie balked at the implementing-an-action-plan step. Even though she had come up with many viable and excellent possible solutions, she simply refused to take that step.

The teacher conferred with her mom and discovered that Julie had always exhibited a lack of confidence about just about everything, and so when she got a less-than-perfect mark, it just confirmed what she believed to be true — that she was a failure. In Julie's case, psychological determinants were at play, and the teacher recommended professional help. The point however is that the child's shaky self-confidence made it impossible for her to follow through with her plan. She was afraid of another failure.

This case is extreme, but many children suffer from low or shaky self-confidence, which significantly reduces their successes as problem-solvers. Hence, it is worthwhile to continually incorporate activities that boost self-confidence into routines. If we want our kids to be lifelong problem-solvers, we have a vested interest in first building their self-esteem. If effective problem-solving is the key to successful adulthood and positive self-esteem is the key to effective problem-solving, our role as educators and caregivers is obvious.

Teachers are well aware of the differences between low and high self-esteem and their effects, but sometimes it's helpful to be reminded. The following list is intended to jog your memory about what self-confident kids can do and to help put you on the lookout for behaviors and activities that induce poor self-concept in your students.

Self-confident kids

- take responsibility
- take risks and chances
- accept failure as part of growth
- remain optimistic most of the time
- act independently
- are able to handle and express positive and negative emotions
- tolerate frustration as a normal part of trying something new
- take pride in their accomplishments
- are willing to try new things
- are willing to help others
- are good problem-solvers

Low self-esteem has been compared to driving through life with a hand brake on. You may want to share this idea with your students, depending on their ages and understanding of driving.

56

There is a wealth of information available for help in this area, so I have chosen to reduce the plethora of ideas to those 10 that I have found the most useful and easiest to incorporate into existing curriculum.

Self-Confidence Promoting Activities

1. Be a great model.
   - Value yourself and your profession.
   - Be positive and encouraging.
   - Be courteous and respectful of others.
   - Be realistic — no one can be positive all the time. Kids need models for all kinds of behavior.
   - Laugh at yourself. Laugh with the kids.
2. Use reality-based praise and criticism.
   - Praise specifically. Say. "I like how you lined up all the numbers," rather than "Good job."
   - Avoid giving praise if something is not praiseworthy. Be honest and provide constructive criticism — criticism about the behavior, not the child: "These answers are incorrect. I think the adding is incorrect." Avoid saying something like "You are wrong" or "You did it incorrectly."
   - Shakespeare reminds us that we cannot all be "masters." Keep that in mind when praising and criticizing. Know when kids are doing the best they can do, and when they are, for whatever reason, sloughing off, and speak accordingly. Think of your words before you utter them, and make sure they mean exactly what you want them to. For example, when you know a child is not doing his best work, a good choice of words would be, "I have seen you do a great job of … in the past. Today this is not your best example. I need a better job from you." On the other hand, when you recognize that a child is doing the best she can, even though it may not be as good as what others are doing, say something like, "I'm so glad you are doing your best at … I can see how hard you are working and that makes me happy. Look at [specifically indicate one part of the task]. See how well you've done this."
3. Set kids up to succeed.
   - Don't treat all kids the same way or expect the same from all.
   - Give responsibilities carefully and expect results.
   - Know the curriculum so that you can make wise decisions based on individual abilities.
   - Have your students assist younger kids with reading or math.
   - Monitor individual work to pick up on little difficulties before they become big, hard-to-solve difficulties.
   - Nothing builds self-esteem like accomplishment. Be sure that every child accomplishes "something" every day, whether it is as small as completing a drawing or as large as completing an essay — it depends entirely on the individual child's strengths with which you are likely well aware. You can ensure individual accomplishments by providing a variety of different tasks each day and quickly checking off a name on a fresh list when a student completes something; near the end of the day, a quick glance will let you know if any student has not accomplished anything. In that instance, provide some small task, such as inviting that student to read a

Teacher Tip

Bring in a colorful poster or overhead of a person and invite discussion about how the person may be feeling. Why? What clues are there? Ask if we can always discern how a person is feeling. When can't we? Kids who tend to keep feelings to themselves may find this discussion enlightening. For homework, they could watch three people to see what non-verbal clues they give as to their feelings.

Teacher Tip

Either bring in a few cartoon strips with the word bubbles whitened out, or ask kids to bring in some, without the words. Invite kids, in pairs, to think of positive self-talk to put in each open word bubble. They will enjoy doing this especially if the cartoon characters are in stressful situations. The exercise also points out to kids that we can talk positively, no matter what the problem.

Teacher Tip

Find pictures, perhaps from kids' readers, magazines, and greeting cards, of people interacting, and have kids think of ways the characters could provide constructive criticism without ridicule or shame. For example, if a picture shows one child hitting another, a constructive criticism might be, "If you keep hitting him, he will be hurt and angry and may even hit you back." The comment provides a reason for stopping the behavior without shaming the child by calling him a bully. For homework, suggest that students keep a record of any time they provide constructive criticism.

short passage or even to take a message to the office, so that she, too, will have achieved something.

4. Acknowledge children.
   - Tell them how or why they are special.
   - Use their names correctly.
   - Greet them as you would a colleague.
   - Acknowledge birthdays.
   - Acknowledge successes, being careful not to always praise the same kids.

5. Encourage personal expression.
   - Model expressing how you feel. Perhaps show extreme displeasure with your body language and face and then discuss this position. You may want to point out that while not all emotions should be expressed, hiding all emotions is not good, either. Sharing anger can allow the other person to rectify the situation and prevent it from growing; sharing embarrassment likely makes an awkward situation worse.
   - Have kids write about their feelings in journals. Suggest that they choose a specific feeling, perhaps one they are experiencing or have recently experienced. They can describe how it felt, what they think caused it, and what, if anything helped it go away.
   - Draw attention to the feelings of protagonists and antagonists in stories.
   - Invite students to produce drawings of feelings.
   - Talk about feelings and how to deal with them often so that kids become aware that everyone has both positive and negative feelings, and has to find ways to deal with them.

6. Teach and give practice in self-talk.
   - Model positive self-talk, for example, say, "I can do this" or "It's okay to lose a game — we're still a good team."
   - Encourage positive self-talk in private journals.
   - Whether positive or negative, point out self-talk in stories, even when it is inferred as opposed to written specifically. Discuss its impact.
   - Share this quotation with kids. Buddha said, "You yourself, as much as anybody in the entire universe, deserves your love and affection."

7. Avoid ridiculing or inducing shame.
   - Think before you speak, especially if angry, tired, or upset.
   - Teach kids to do this, too.
   - If you do slip, be sure to apologize, one on one, as soon as possible after the slip.

8. Teach "tease tolerance."
   - Practice dealing with situations where a child is the brunt of teasing (or even bullying, although that can become a bigger issue).
   - Self-disclose a time when peers teased you — situation can be real or imaginary — and share how you felt and what you did.
   - Help the kids to figure out ways to deal with a tease situation other than by getting upset, crying, tattling, or hitting. Point out that sometimes a teaser isn't trying to hurt them and that they may have taken words the wrong way. Asking for clarification can be a good, perhaps the only needed, step. Have the class brainstorm for ways to react positively to teasing. These may include making eye contact and calmly saying "stop teasing me," walking away, ignoring the teaser and doing something else, and shrugging or smiling as if not bothered. In the case of more severe teasing bordering on bullying, recommend that kids follow these steps.

This simple technique for dealing with teasers works miraculously. I have witnessed it many times. Encourage kids to practice it in groups and to memorize the words so that they come easily in an otherwise tense situation. You can point out that this is a way of problem-solving.

1. Stop and look at the "teaser."
2. In a calm voice, say, "If you're trying to hurt me, you've succeeded. What you [said, did, implied …] really hurt."
3. Immediately turn and walk away.
4. Say nothing more.

Point out how, by doing this, the person being bothered gains the upper hand. The "teaser," or bully, obviously wanted to "hurt" so when the teased person acknowledges that, there is no longer a reason to keep up the hurtful behavior.

9. Teach independence.
   • Teach problem-solving skills.
   • Practice problem-solving skills.
   • Point out where others have successfully used problem-solving skills.
   • Avoid solving problems for them (except for the big ones).
   • Set clear expectations and classroom rules.
   • Expect that kids will follow your rules and meet your expectations.
   • Allow free time (unless it is being abused).
10. Be there.
   • One important way to improve a child's self-esteem is to "be there." I don't mean being there every minute of every day, but giving your full, undivided attention when you are with the child. Give.
   • Be true to your word. Children need to know where you are, for how long, when you'll be in class or at home, at least as much as possible. By committing to this, you let them know that they are important enough for you to feel obligated to follow through on what you have said or promised.
   • Give time to listen. Be available particularly in times of distress and stress. Children will understand that since you care enough to be there, they must be special, important, and worthy of your time.

Rather than a game for this chapter, I recommend sharing a good children's picture book that illustrates the benefits of self-confidence. My favorite, which works with all ages, is *Knots on a Counting Rope* by Bill Martin Jr., John Archambault, Ted Rand, and Bill Martin. In this colorful tale, a young First Nations lad, born blind, has the confidence to ride a pony in a race.

Self-Confidence and Problem-Solving

Self-confidence and problem-solving work well together. One without the other is probably almost impossible. Which comes first? That's like the chicken-and-the-egg quandary. I can't give you an answer. They are equally important. Deal with them simultaneously, in an ongoing, consistent, patient way, and the results will be impressive.

8 Troubleshooting and Consolidating Problem-Solving

Madeline was a problem-solver. All the other kids knew that; they went to her with their problems and she solved them. She had been dubbed the "Official Grade 8 Teen Shrink." She loved the title.

However, when it came to solving her own problems, Madeline was less than stellar.

Her teacher had witnessed this on several occasions and shared the following episode with me. Apparently in her role of "Teen Shrink," Madeline was the recipient of some sensitive information about a peer's family, confided to her by the peer, whom I'll call Kelly, along with a firm promise of secrecy. The news was too juicy for Madeline to keep to herself, though, and she shared it with her best friend. You know what happens when teens "share a secret" — it travels with light speed accelerated even more by modern technology. The result? Madeline had to face Kelly and her parents, own up, apologize, and accept responsibility for the mess she had created.

Madeline's teacher encouraged her to follow the Five-Step Problem-Solving Plan, which she did quickly and expertly; however, when it came to taking the first step in her action plan, Madeline simply couldn't, wouldn't do it. The teacher tried various ways to get her to take the step, but failed. The story didn't end there, but for purposes of this book, it does.

The point? Madeline's teacher had to troubleshoot. She had a student who was "stuck" and wanted to find a way to nudge her ahead.

Nowadays everything comes with a troubleshooting manual. I'm a firm believer in the power of the positive attitude, and hence, having no need for troubleshooting, but experience has shown me that when working with kids, things seldom go as planned. So, here are some possible "what-ifs ..." and some possible "then try's...."

All the What-ifs

What if every small situation seems like a problem? This perspective points to the idea of *choice* versus *problem*. If a student is in my-whole-life-is-a-problem mode (often typical of adolescents), it's time for the daily dose of choices. The student makes a minute-by-minute record of every problem she thinks she's encountered in a single day.

- She carries a notebook and jots down every single thing that "troubles" her.
- At the end of a day, she reviews her jottings and highlights the problems according to levels. Colors might be red for really serious Level 3, blue for Level 2, and yellow for Level 1. If something was a "choice," as in deter-

Teacher Tip
You can use this same technique with an entire class. Together, create an imaginary day for an imaginary person, and have the students suggest realistic, possible problems that could occur, for example, missing the school bus; then, highlight together and determine which problems would be genuine problems and which would be just annoyances. This is a good time to point out how many decisions each child makes in a day and how "special" that ability is.

mining what to wear, it shouldn't be highlighted. In all likelihood, there won't be any red highlights.

- The next step is to try to justify any highlighted "problems." Most often, the student will decide many of them weren't problems, after all.

What if a problem-solver can't seem to take the first step in an action plan? There could be many reasons, beyond the scope of this book, for this behavior. However, as educators, we should provide the impetus necessary for the reluctant child to move ahead with a planned solution. Ideas to help the reluctant starter include these:

Teacher Tip

There are many possible reasons for being unable to take that first step. Common ones with which you may be able to help are fear, lack of confidence, and lack of information or personal skill.
Fear: Talk about it, using the "what would be the worst possible outcome and can you handle it? approach.
Confidence: Use any of the Chapter 7 suggestions.
Lack of information or skill: Provide information where possible and help the student either acquire the necessary skill (e.g., rewriting an essay using correct form) or change the possible solution (e.g., writing just a paragraph instead of the entire essay).

- Encourage him to challenge himself: "Give yourself a time to start, then challenge yourself to do better."
- Reinforce even a hesitant start, using specific, realistic reinforcement: "I like that you've taken the first step because now you will be ready to…."
- Invite the child to put the action plan away for a full day, even a week, and tell her not to think about it during that time. (This is like the "Don't think of purple cows" idea, where the mere suggestion makes it impossible not to think about.) At the end of the time, revisit the action plan together. The child is usually ready to start.
- One on one, discuss possible reasons why the child seems unable to start. Avoid asking, "Why …?" The child probably doesn't know why and this approach may further alienate him. Instead, say something like "I'm confused about this difficulty in getting started. Let's think together to figure out what might be in your way."
- As a last resort, you may need to get assistance from a school counsellor. A stubborn inability to get started may be a symptom of a deeper problem.

What if a child needs help with his problem, but keeps insisting he can do it alone?

- Reassure the child sincerely that you understand his desire to do it alone and respect it, but that you would feel better if he'd allow you to help. By using an "I" statement ("I would feel better …"), you avoid possible arguments.
- Self-disclose a time when you needed help as a way to show that it's okay to get help.
- If the child is still reticent, ask if he would accept help from someone else in the school. (Sometimes, gender or even age makes a difference.)
- If the child still refuses help, and you are concerned about it, then call the parents. Before doing so, however, tell the child that you plan to do this and explain why.

What if a child always insists she has no problems, even after discussion about all the little problems people face daily?
If you are aware of a problem the child has (e.g., a gravely ill parent), but is refusing to acknowledge, consider that the child might be in denial, and deal with it as denial, as this can be a cause for concern. Here are a few possible ways to handle this:

- If the child cannot identify a personal problem, try giving her a "fake" one to work through, telling her that it will given her a chance to learn the strategies. Sometimes, by working with a pretend problem, real problems come to light.
- In a one-on-one situation, ask whether there is a particular problem, but avoid making specific suggestions, such as "I hear that your mother is ill…."
- Discuss your concerns with a school counsellor.

Sometimes children don't think of little annoyances, like being unable to find a backpack, or the need for quick decision making, such as what to take for lunch, as problems. It's good when kids see such matters as less than disturbing; however, if you want the child to look more closely at coping strategies, including those that enable him to make good, rapid decisions, try the following.

- In a one-on-one situation, suggest mini-problems that the child probably encountered that morning (e.g., chose what to wear, what to eat, who to talk to in the schoolyard). If the child does not accept that any of these were mini-problems, accept his reasoning, but be alert to his future moods and behaviors.
- Remember that it is always possible the child doesn't have any problem worth mentioning.

What if a child can handle "big" problems, but seems completely flustered by little ones?
Surprisingly, this is not uncommon. Solving and managing a big problem takes full focus and concentration, and some kids excel at this. They can attack a big problem — a Level 2 or even a Level 3 problem — head-on, and are usually successful with their action plans. When faced with daily mini-problems, however, they become frustrated, even anxious. I'm sure there is a psychological explanation for this, but as far as educators are concerned, we want to know what to do about it.

- One-on-one discussion is the first step. Ask pertinent questions such as "What upset you about …?" or "Why did … make you angry?"
- Help the child to see that these minor inconveniences are a part of life. Point out how many there are in a normal day and how exhausting it would be to react with frustration and anger to each one. Use humor, self-disclose (even if you have to make it up), and use exaggeration, as in "What would happen if I broke my pencil in class and then started screaming and stamping my feet and pounding the board?"
- Help the child make a list of all the mini-problems faced that day and try to bring humor to each one. The act of identifying them is often enough to calm the child and avoid future overreactions.

What if a child continually makes poor choices in Step 3 of problem-solving and ends up failing or sabotaging himself?
If a child is constantly making poor choices, it's time to step in and go over the following questions together.

- "How many times have you tried to solve this problem?" Sometimes, drawing attention to this helps a child make better choices the next time.
- "Let's look at your attempted solutions. Why do you think they failed? What can we do differently?" Look for a pattern of incorrect or inappropriate behavior. If you can't find one, suggest a completely different approach. For example, if a student is trying, but failing to improve marks by adding more and more time to at-home study, recommend an after-school study class, tutor, or "study buddy," an older student who is proficient in the student's weak area and willing to commit time to helping.
- "How many other possibilities did you come up with when at Step 2? Would any of these ideas work? If not, let's brainstorm again, and think of more alternatives."

What if a child skips the brainstorming step (Step 2) and always goes right to identifying a possible solution?
No matter how well the child manages to choose a solution, it is still a good idea to reinforce the generation of ideas, as in Step 2, "for future use."

- Provide specific positive reinforcement for the child's ability to arrive at successful solutions without acting on Step 2, but challenge her to deal with another problem (real or imaginary) and include the step. For example, a student forgets her gym shorts and jumps to the solution of not attending gym. Brainstorming might yield borrowing shorts; asking the gym teacher if she could participate, even partially, in what she was wearing; getting permission to go home for the shorts; and calling her mom or dad to bring the shorts to school.
- Point out why the step is necessary. (There may be a time when it's difficult to make the right choice; the student may be missing a possibility that is equally as good as her choice, and maybe even better.)
- Play brainstorming games in class and make a point of reinforcing her contributions.

What if a child always "adapts" — that is, uses the Peregrine Falcon approach — falling back on familiar tactics even when a more creative, new approach would be better?
We all adopt this safety measure, especially when constrained by time or faced with a number of problems at once. There is nothing wrong with using the adaptive approach; however, as educators we want our students to experience all approaches so that their adult lives will be full and productive. We want them to feel confident about solving problems when they don't have any tested solutions to fall back on.

- Provide sincere, positive reinforcement for the child's ability to call on familiar measures.
- Remind the student of the other approaches — specifically, of the Chickadee approach — and challenge him to find a new solution to an already solved problem.
- Pair this child with another, and present them with a real or an imaginary problem to solve, one that will require innovative thinking. For example, a student named Gwen always seemed to break the laces on her gym

shoes. She solved the problem by tying them together, and this worked, until the day when they were too short to tie. Her teacher encouraged her to think like a Chickadee, and invited Gwen's buddy to help. Together, they came up with the idea of pulling an unneeded tie from the waist of Gwen's sweatpants, and using it as a shoelace.

Examples of imaginary problems that usually gets kids thinking innovatively:

1. You and a friend are lost in a dense forest. You have no cellphone service, flares, or matches. What do you do?
2. Somehow, you and a friend got locked in the school after dark and can't get out. You lack access to phones, computers, and technology of any kind; you cannot control the electrical light sources, and you have no flashlights. Your parents will be worried. What do you do?

What about the Little "Annoying" Problems?

I have alluded to these at other places in this book, but feel it's important to give them a place of their own, if only because so many children — and adults, unfortunately — tend to overreact, to be "Peacocks" at the slightest provocation. As educators we are responsible for helping our charges learn how to deal with what I call Level 1 problems quickly, effortlessly, and effectively.

I have found it effective to draw attention to such problems and to invite kids to look carefully at how many of these they solve daily, without even thinking about it. Prompt students to list all the "annoyances," or Level 1 problems, they have already had that day. You can help them by self-disclosing: the alarm didn't go off, you spilled your coffee, you stubbed your toe, you dropped your keys twice. Then, ask them to each create a poster or visual — a "road map" — showing all their little "victories." Refer to each solved situation as a mini-victory — doing this gets kids thinking of their own skills as problem-solvers. It also lets them see the difference between problems and annoyances, or Level 2 and Level 1 problems.

Once students can discuss freely the idea of mini-problems and can differentiate between gravity of problems, a few key ideas can be shared. First, although an awareness of the Five-Step Problem-Solving approach is important, it is not used for all these mini-problems. What is used is common sense. Most people do a quick mental run-through of *most* of the five steps even with the smallest of problems.

Example: Spilled grape juice
Quick thoughts: Leave it, wipe it up, swear, shout, sigh, pour more, don't pour more

It takes just seconds before someone acts on a thought. But even with this instant approach, a few common factors are at play, and we can share these with our students. Of course, these ideas apply to any and all problem-solving situations, but remembering them is especially important when solving on the run.

Although the poster-making activity is better done independently, if the children are very young or reluctant, allow them to work with a partner to create a poster of shared victories.

Quick solution ideas

- Use common sense.
- Think "safety first."
- Ask yourself what someone you admire would do in this situation.
- Consider the worst possible scenario. (If you can handle that, and we almost always can, then nothing else seems so bad.)

Sample problem, student perspective: Dropped homework in the mud on the way to school

- Common sense: Pick it up and wipe off as well as I can.
- Safety first: If I dropped it on a street, I'd look for vehicles before bending to pick it up.
- What my dad would do: He'd go right to the teacher and explain.
- Worst possible scenario: Teacher throws a fit and kicks me out. No problem — I can deal with that. I'd just go home.

Dealing with Mistakes in Solving Everyday "Little" Problems

Once kids can see that little mistakes, or mini-problem–solving errors, aren't always avoidable, they can be taught effective ways of dealing with them. These ways include the following:

Teacher Tip
Present the class with a situation in which a mistake has been made. You could present a mistake as your own or as that of an unknown person. An example: "I forgot to hand in my report card marks before the deadline." Go through the suggestions outlined at right, and then ask students how they would deal with the specific mistake.

On the other hand, here are a few key phrases to be aware of and avoid:
Yes, but …
I couldn't help it.
Not my fault!
I was just following orders.

- Using humor: Encourage students to try to keep a sense of humor. It really helps with daily problem-solving if they can see the lighter side of it. It takes no longer to feel silly and laugh at what they have done than it does to be angry and snarl. Snarling also takes more energy. This idea is not new, but it is worth remembering.
- Being positive: Here's that neutral mood again (see page 17). Even if students are not feeling "neutral" when the little annoyance appears, the best thing for them to do is to strive for a neutral frame of mind until the mini-problem is solved. If they let it, an already sour mood can easily become sourer yet. On the other hand, if a student deals with an irritation in a neutral manner, it will be so much better for all concerned. Here are a few key phrases worth memorizing:

  I have a choice
  I choose to be neutral
  I am smart
  I am responsible and I can take care of …
  I can make this okay

- Accepting responsibility when necessary: Students need to be able to say "my fault" — kids like the phrase "my bad." It doesn't matter how they say it; they just have to say it, to practice it. Saying it to themselves when they have done something stupid forces them to solve a little problem. By accepting responsibility for the little things, they will find it easier to accept responsibility for the big faults, too.
- Accepting consequences: It's not enough for kids to accept responsibility. They must also accept consequences. Many kids find this tough. They are

quick to say, "My bad, I'm sorry." But they are less ready to accept what follows.

- Avoiding the same mistakes again: This is the learning, the growing part of making mistakes, but many of us have a lot of trouble with it. Every tiny problem solved should mean that the same problem doesn't crop up again. For example, if coffee was set too close to the edge of a table and spilled, it shouldn't ever be set too close again. I guess "should" is the key word. Suffice to say, it's good to remind our kids that we can learn from mistakes and that problems are opportunities for learning.

Why Problem-Solving Strategies May Fail

We are all aware of problems that just couldn't be solved or problems that we believed we'd solved, only to find out later they still existed. Similarly, there are those annoying problems for which we simply can't find suitable solutions. Learning when to stop, when to retreat, and when to accept the situation for what it is and move on without a "happy ending" is all a part of growing up and becoming independent.

There are, however, key reasons why a problem may become unsolvable, including these:

- failing to properly identify the problem
- addressing just one "symptom" of a problem and not the whole problem (Problem — A student fails math. This could be a symptom of the problem she doesn't know how to study.)
- using vague or stereotpical descriptors when defining a problem (That person is too old to …)
- confusing fact and opinion (Problem — A student thinks, "My mark is not good enough" (opinion); she could think, "My mark needs to be 10 percent higher to reach a passing grade" (fact).
- failing to consider many or at least several solutions before implementing a plan of action
- failing to use more than one approach to problem-solving — sometimes, the Robin approach works better than the Peregrine Falcon approach

Class Problem-Solving — Finishing with Flare

Everything a teacher teaches should be finished with flare. In this way the students will remember at least the key elements. To finish any lessons on problem-solving, it is recommended to involve the students in a realistic problem-solving situation.

I suggest a group problem assignment. The assignment can reflect any theme then being studied; however, the assignment will stop, for the problem-solving component anyway, when the action plan has been described. In other words, the students are responsible for all steps of the problem-solving process, for recording in writing each step, and for creating an action plan. A curriculum example comes from the Grade 4 "Waste and Our World" component, in which students study consumption and waste, identify wastes produced within their community, and learn about disposal methods. They could deal with the

Teacher Tip
One excellent idea I saw implemented in a Grade 5 classroom was that of a Mistakes journal, a small writing book in which each child entered mistakes made, together with the dates and the subsequent learning. Example: *Tues., May 2, 09 Forgot lunch; learned to put it by my backpack when I get up each morning.* Students took a few minutes at the end of each day to write in their books. The journals were not evaluated although the teacher's comments could be seen throughout.

problem that their school needs to find a way to cut down on paper use, even though most of the paper used is recycled.

The entire project becomes much more meaningful, though, if, as a class, you can carry out the action plan together. In order to do this, the original problem must be relevant, realistic, and solvable. Here are some examples.

Possible problems for a class to deal with

- Class needs money for a field trip which costs more than individuals can afford. They need to find a way to earn $60.
- Class wants to experiment with a recess lemonade stand. They have neither the materials to build the stand nor the products to make the lemonade, and have only a $25 float.
- Class wants to have a June picnic, but it's already May and they didn't reserve a site at the local park.
- Schoolyard and surrounding areas need to be cleaned up, but no one wants to do it. The students must create a plan in which every class shares equally in the cleanup and receives some sort of reward or recognition for their efforts.
- Class wants to welcome all new Grade 1 students to the school, but wants to do something more creative than the usual meet and greet event.
- Class wants to plan something for Open House Night. They need to decide on a few good ideas that will demonstrate their learning and accomplishments, and create an action plan to get them ready to present in time.
- Class wants to volunteer in the community, but they don't specifically know how, when, or where.

As you can see, any of these problems could be taken through the Five-Step Problem-Solving Plan and even put into action, at the teacher's discretion.

Putting Problem-Solving in Its Place

I have offered many ideas about problem-solving in this book, and since my goal is to help teachers use the information in their classes, I have included a unit plan in Appendix C. In addition, here is a quick review of the key elements of problem-solving, to serve as a reminder, a teaching guide, or the text of a wall chart for students.

Problem-Solving at a Glance

1. Establish problem reality. Assess whether you really have a problem.

2. Think about what's in the way of your solving the problem. Is fear involved? Is a single strong emotion making it hard for you to solve the problem?

3. Determine which overall approach to take to problem-solving. Of the seven bird approaches you have learned about, which do you usually use? Will it work here?

4. Consider these steps to approaching a specific problem:
   • adopting a neutral mind (or, seeing the problem positively)
   • determining the problem's seriousness
   • using the Five-Step Plan

5. Follow the Five-Step Problem-Solving Plan.

   1) State the problem clearly. Define it by carefully stating it or writing it down.

   2) Determine the possible choices — come up with many possible solutions.

   3) Assess the possibilities and make one or two choices.

   4) Devise and implement an action plan, being sure to write it out and to include a goal and a timeline.

   5) Evaluate the success of the problem-solving. Did the solution work? Why or why not?

9 Helping Parents Deal with Children's Problem-Solving

I applaud the mother who allows her four-year-old to problem-solve whenever and wherever possible. There are no rules stating that children have to be a certain age to start taking responsibility for their actions, although many parents might argue that. The younger a child is exposed to problem-solving, the more likely he or she will be to develop this invaluable life skill. We do kids no favors by solving all their problems for them, even though — and I'm the first to admit to this — it is usually easier and quicker to take charge ourselves.

A front-page newspaper story told about a young child who came across her mom lying unconscious on the kitchen floor and immediately called 911. When the medics arrived, the child was sitting calmly beside her mom, stroking the woman's head. She had covered her mom with a blanket. She was four years old and already a master problem-solver!

While visiting a friend who was babysitting her four-year-old granddaughter, I was surprised to note that the little girl, Jessie, was dressed in the strangest of outfits. Not even her shoes matched, and she looked quite ridiculous. Curious, since I knew that the child's mother was a fastidious dresser and Jessie had plenty of outfits, I questioned my friend.

"Oh that," she said, raising her eyebrows. "Yes, well, Jacquie [Jessie's mom] is a new-age mom; she believes in letting Jess problem-solve as much as possible. What to wear is Jessie's problem. She solves it, and unless she picks something *weather* inappropriate, Jacquie accepts the choice. Things have changed since we had kids, haven't they?"

Consider another incident, this one involving school. A Grade 4 student, Susie, missed a field trip because she hadn't brought the required $10 for the bus fare. Now, as a rule, the school will cover for students unable to bring the fee, but in this case, a note from Susie's mom advised us that we couldn't step in and that Susie would have to miss this trip. A very unhappy Susie, who was dispatched to another class for the day, finally told the story. Her folks had agreed to supply $8 toward the cost of the ticket, if Susie worked to earn the other $2 herself. Susie's parents could afford the entire amount, but they wanted Susie to take responsibility for her share. It was her problem. Apparently Susie challenged her parents by saying that it wasn't fair and that she didn't know how to get the money, to which her dad calmly responded that she had a problem to solve before the field trip. Susie, however, chose not to solve it. She approached the problem as an Ostrich, assuming her parents would come through at the last minute. They didn't, and Susie had to deal with the consequences of her Ostrich behavior. I'm happy to say that when the next field trip arrived, Susie was one of the first to hand in the necessary fee and was quick to explain how she had "solved the problem." I was proud of her, but even prouder of her parents.

Another incident concerns a fear-based problem. A Grade 3 student, a sweet, small child with huge eyes, was terrified of dogs. Her single mom confessed she had no idea why Hanna felt that way and worried about its ultimate effect on her only child. As luck would have it, a dog owner moved into a house on Hanna's street, and Hanna had to walk past the house daily on her way to and from school. The dog was always chained in the yard, and as the mom was told by the owner, perfectly harmless, but it barked loud and furiously when anyone walked past.

After consultation, Hanna's mom explained to Hanna that it was her problem to solve, but that she'd be there to help her find a solution. Together, they worked through the problem, from defining it (according to the steps in Chapter 2), brainstorming possibilities (as outlined in Chapter 3), and then evaluating the list of suggestions (as in Chapter 4). I won't go into all the possibilities, but Hanna and her mom ended up with the following two possible solutions: take a different route home or find a friend to walk with. Hanna chose the first alternative. She found an alley that kept her away from the dog, and although it took her a little longer to get home, she did it with a huge smile.

In all three incidents where a parent played an active role, the parents allowed and encouraged the children to problem-solve. It wasn't easy for them. Jacquie admitted that she would much rather have seen her daughter dressed in matching, fashionable outfits. Susie's parents explained how terrible they felt about her missing the field trip. Hanna's mom said she waited in a state of high tension while her daughter walked the alley that first day. These parents all showed confidence in and respect for their children.

Most parents will readily agree that they love their children and want the best for them, but sometimes they need a little help in understanding just what doing the best for their kids involves. They don't want their children to suffer, to feel pain, to be worried or upset. They don't want to watch them fail. It's tough to witness their disappointments and their fears. They want to protect them from experiencing the less-than-fun sides of life. They may want to rush to the rescue, but they should not.

Adult Behaviors Not Conducive to Children's Problem-Solving

Have you ever observed any parents trying to problem-solve for their children in the following negative ways? Or, perhaps, you have engaged in some of these behaviors yourself. Watch out for these parental behaviors:

- making excuses for children's inappropriate behavior(s)
- telling little white lies to "cover" for them
- explaining to you as the teacher why an assignment is (late, undone, poorly done …)
- giving them money to "get them out of their hair"
- giving in when the children whine and beg to do something they don't want them to do
- helping them cover up a mistake they've made
- stating a "consequence," but then backing down because they don't want to upset the child
- lying to someone on the phone because their child doesn't want to speak to that person
- using money or gifts or bribery to deal with kids' negative feelings due to a situation the parents have created (for example, providing peace offers, such as money, to supposedly show or even buy love when they are seldom home due to work and other responsibilities, and kids consequently feel somewhat left out)

But — and this *but* is huge — there are situations where, as adults, parents and teachers *must* solve kids' problems. These are the life-threatening situations, the

Level 3 problems, the overwhelming problems that are far too big, too dangerous for anyone to face on their own, let alone a child. In this chapter, I am not talking about those life-threatening problems (see Chapter 6). This chapter is designed to help parents make their behaviors and interventions more child-as-problem-solver friendly, and it offers guidance to teachers on how to deal effectively with parents and support them in their efforts with their offspring.

Understanding Why Parents Solve Their Children's Problems

The initial step is to encourage parents to do a self-analysis to determine why they may be prone to solving the child's problems. Their immediate thought is usually "to help my child," but more often than not, there's another reason or possibly several other reasons.

Why parents typically solve their children's problems

The following reasons are presented from a parental perspective.

- It will take too long for him to solve it. I'm busy, I have to get going, I need it fixed, dealt with, handled, solved, right now.
- I can't deal with my child's emotional reactions to this problem when he or she fails to solve it correctly. I'm too (tired, anxious, overworked …) right now.
- I don't want my child to be mad at me. I don't have the energy to handle rejection, anger, accusations, or the silent treatment. I need a little love right now.
- It's just easier to "do it myself." The learning curve for my child is too great for me to deal with at this time. I'll solve the problem and be done with it.
- I don't trust my child to handle, deal with, solve it properly, and the ramifications of an incorrect solution will fall back on me.
- I don't want others to think I'm not a good parent and if my child can't fix the situation, that's what will happen. My reputation is at stake.
- I'm too tired, overworked, busy, strung out to help my child think of possible solutions and then probably execute them incorrectly anyway. I'll just take care of things.
- My child can't possibly handle this mess, so I'll do it.

These ideas can be summed up by the sentence "I don't want my child to suffer." I can honestly say that I have been guilty of every one of them in the raising of my own children. I wish that I had known then what I know now. I wish that someone had told me that the only way to teach a child to problem-solve in adulthood is to coach him to problem-solve as a child. To let him falter, fall, fail. The only way to promote independent learning, self-sufficiency, and self-confidence is to promote early problem-solving with its consequent learning and growth.

A balance between showing care and promoting independence

Teachers can help parents find the balance between not wanting to see their children come to harm (consequently solving every problem for them) and promoting independence (encouraging children's problem-solving skills). On the one hand, concerned and loving adults are natural "rescuers" who jump in and "save" children, who solve problems so that "they won't get hurt." On the other hand — and here, teachers may have a special role to play — adults may have other reasons for intervening that could be brought to the fore.

Could it be that the adult doesn't want to deal with the repercussions that will follow if the "child" doesn't solve the problem correctly? Consider the following. A mother of a 14-year-old relents and gives her daughter money to go shopping with friends, even though the girl has already spent her allowance and hasn't saved any money from babysitting. The mother knew that if her daughter didn't go, the girl would sulk all weekend, making home life miserable for everyone. The mom cuts her losses and gives the money. From this cut-my-losses-to-maintain-personal-sanity situation, the child doesn't learn to problem-solve and ends up expecting the parent to bail her out, even for the rest of her life.

In the belief of helping the child, parents far too often help themselves and ultimately hurt the child. Without realizing what they are doing, parents may create a situation of co-dependency in which the child forever expects Mom or Dad to solve all the problems, especially those related to finances. Mom and Dad do exactly that, usually at their own financial, emotional, and even physical expense. They may later find that they are expected to step in even when their child becomes an adult.

Sometimes it just takes another concerned adult, the teacher, to draw this issue to parents' attention and help them make changes that will benefit everyone. Teachers can help parents understand the dynamics of this type of situation by mentioning them at parent–teacher conferences or by sending home relevant information throughout the year.

Teachers fall prey to this behavior too when they inadvertently solve children's problems with the intent of "getting them under control so that the lesson can be continued." Doing this is sometimes necessary, of course, but it's important to be aware of when it happens, how often it happens, and with whom it happens. If it's always the same child the teacher is solving problems for, it's time to examine the situation more closely.

In considering their responsibility to the whole child, teachers can play a role in preventing situations where parents, or even teachers, always solve a child's problems. As teachers see need and opportunity, they may be able to guide parents by suggesting ways to change their dealings with their children. It takes courage on the part of the adult — more courage than to give in and "solve" a problem. The situation can be considered a problem that needs to be solved. Using the rational problem-solving steps delineated in previous chapters, I have outlined some ideas for teachers to share as they find appropriate.

The line master outlines wonderful outcomes from simply doing "nothing." Remember, though, that I am not referring to the huge life-threatening problems that, unfortunately, children sometimes encounter. In those instances, caring significant adults should jump right in and rescue, rescue, rescue. It is all the other problems, large and small, that we as adults somehow feel responsible for. It's time to stop; it's time to say, "No" and "This is your problem to solve, but I'll be here to help."

One of the first anecdotes in this book, the tale of John and his expensive broken tool, is a perfect example of an adult still turning to a parent for problem-solving.

72

The Adult Problem-Solve Situation

Everyone faces problems and for adults, including parents, some of those problems involve how they interact with their children when their children have problems to deal with. Here is one workable approach to that situation.

Problem definition: *In what ways can I stop solving all problems just to "keep the peace"?*

Possibilities: *Just say no, laugh, just do it, get father/mother to help, ask a school counsellor, get counselling for me, see a psychiatrist*

Evaluation of ideas and choice: *Choose "just say no."*

Action plan: *Practice saying: "No, this is your problem to solve. I will help, but I won't solve it for you." Use the five steps of the action plan listed below.*

Evaluation: *Next problem: How did the practiced sentence work?*

Adult Action Plan

1. First, recognize if you are constantly "bailing out" your offspring. Stop, take stock, and understand that you are solving the problem for yourself, not your child.
2. Tell your child that you won't continue this behavior. Explain why. (You are impeding your child's ability to problem-solve and hence be happily independent.)
3. Stick to your word, which will be tough. There may be many nasty repercussions the first few times you refuse to solve a problem, but be vigilant. Remind yourself that you are stronger than the child (no matter what age that child may be) and that you are doing what's best.
4. Reward yourself when you do stick to your word. The nature of the reward is unimportant. It's the idea that counts.
5. If you slip, don't slide. Start again with the next problem and explain to your child that you "were wrong by …," but that you intend to return to your original stand.

The Benefits of Stepping Back

There are multiple benefits when adults, be they parents, coaches, mentors, teachers, or any other significant persons, courageously step back and offer only support and encouragement instead of solving a child's problems. If you do so, you can increase the child's self-confidence, independence, problem-solving skills, growth and learning, and respect for you. You can also decrease adult angst and even anger at always "being expected to save," possible financial drain, and adult headaches or other physical symptoms that tend to appear during a child's crisis. You will find that you have greater respect for the child. There should also be less tension between spouses, between peers, between other students — a lot of tension is caused by one or the other being a more frequent problem-solver. Think how great all of that could make you and the child feel.

Promoting Children's Independent Problem-Solving

A child who is accustomed to parents bailing her out of dilemmas may react loudly, angrily, and with downright indignation. The common phrases are "Why are you doing this to me?" and "That's not fair" and "You always helped before."

If parents express an interest in changing the way they approach their children's problems, the teacher can be a source of valuable, helpful information by offering support and reinforcing their desire to take positive action. She can then point out that if the children have never had to problem-solve because parents have always done it for them, it's possible they, at least initially, won't be able to manage without parental input. A child caught in this situation, one who has never been responsible for his or her problems or their consequences, will probably stumble quickly when left to problem-solve alone. The teacher can help parents deal with this situation and become more effective at teaching their children to solve problems on their own.

There are some useful responses to the familiar offended pleas that teachers may want to share with families. Many of these are outlined on "How to Respond to Children's Objections." Teachers may find it appropriate to suggest that if any (or all) of the ideas on work, then parents may want to commit them to heart so that they will be ready for the next situation with their children. However, making the transition from solving children's problems to insisting that they take responsibility for doing so isn't easy.

What if the child doesn't manage? What if he or she fails? Adults should support, guide, and offer suggestions, but never go back on their word and solve the problem or bail the child out. One story I was told really pounds this idea home, and I thank the brave mother who shared it with me.

John was what teachers dubbed a "troubled teen." He was always doing something wrong. He had hooked up with a gang that was heavily into shoplifting. His mom found out and warned him that it was illegal, wrong, stupid, and that he'd end up in jail and she wouldn't bail him out. He joked that she'd never let him spend a night in "the slammer" at the age of 16. She again asked him to curb his ways. He didn't. He ended up in jail, just as the mother had predicted. The "jail" really was a jail: a single barred room where all the "criminals" (drunks, vagrants, vandals, and even one repeat felon) were incarcerated.

John found himself in a very crowded, very uncomfortable, very frightening place. He immediately called his mom. She spoke to the sheriff who said she could bail him out or he could stay for the 48 hours the constable had recommended and then face the charge in court sometime in the future. True to her word, she didn't bail her son out. She did, however, go with him to court and help him work out an action plan to deal with the community hours he received, as well as one to deal with his illegal behaviors. She said that the 48 hours when John was in jail were the worst hours of her life. She said she never stopped crying and worrying about him. But she had courage. She was strong.

And the end of the story? Today, years later, John is a prestigious lawyer who often does pro bono work with troubled teens.

Problem-solving starts at home with the parent(s) and then moves to teachers at school. Parental love, understanding, support, and courage are the cornerstones of a child's future as a successful problem-solver. The caring, understanding, support, and understanding of teachers underpin continued growth and self-confidence in this area.

How to Respond to Children's Objections

Below are some probable responses that children may make when adults first expect them to accept responsibility for their own problems, together with possible ways for the adults to react effectively.

- *Why …?* I have decided it's time for you to solve your own problems. I have been wrong to always do that for you.
- *Not fair …* Actually, it is fair. I have been unfair to you in the past by not giving you a chance to solve your own problems and grow.
- *Always helped before …* I was wrong. I now know that it's best for you to do this on your own so that you can grow.
- *I hate you for not …* I can see that you're angry and you have a right to be. Not because I won't help now, but because I have done that far too long and not respected you enough to let you solve your own problems.
- *I'll get … to help …* That would be foolish because it would only show me that you are too immature to handle your own problems. I respect you too much to think that you will do that.
- *Mom, PULEEEZE …* I know you want me to help, but because I love you I will not help you solve your problem.

Of course, there are many more responses kids come up with when they are suddenly deprived of support they have always had. The key is to truthfully explain that you have been wrong in the past and that you will not step in this time because you love them. Yes, this sounds a lot like "It hurts me more than it hurts you," but that's what you are saying and doing, and it *is* going to hurt you until you see just how well your child can manage.

10 Games That Promote Problem-Solving Skills

A note about "games": With today's craze for electronic toys, you might be surprised at the effectiveness of these simple, almost old-fashioned games. They are really more like educational tasks than games, but introducing them as games, either competitive or cooperative, helps kids come to love them. The games enrich the learning process, enhance understanding, and reinforce concepts taught by you.

The games featured in this chapter all have a connection to different aspects of rational problem-solving and help students to practice specific skills required to successfully solve problems, in a fun and non-threatening manner. For instance, Brain Blast explores and encourages divergent thinking, necessary for Step 2, Determining Possibilities and Choices. Similarly, by participating in Name Game, students will practice using precise words to concisely define a situation, much as they have to do in defining a problem.

The games are presented in alphabetical order. Each involves minimum preparation and equipment.

BRAIN BLAST

A competitive game of both chance and divergent thinking, Brain Blast encourages groups to cooperate to come up with as many ideas related to a specific topic as possible. Although this game works well when teaching about creative ideas for solving problems, it is also useful as a prelude to many creative writing projects.

Players:
Whole class as two teams

Materials:
- One large die, which is available from dollar stores, games stores, seniors' stores (You can use a small one, but the large ones are better.)
- List of "umbrella topics" or general theme words (See below; it is a good idea to use concepts being studied in class already as it encourages recall of facts.)

Rules:
According to the number rolled on the die, teams provide the number of correct word or phrase choices that fit the provided theme word. (Example: Number 2 equals two words provided.) If the two words are correct, the team earns two points. If the team can come up with only one correct answer, the team earns one point. The teacher keeps track of scores in some manner visible to students, perhaps using a chalkboard, whiteboard, or overhead. Usually, the first team to reach "20" wins. Sometimes, though, determining the winner is up to the teacher. If both teams end up tied at the end of a designated playing time, the

Creative thinking is encouraged. Points may be awarded for fantasy or made-up words as long as kids can justify them. For example, "plabitat" could mean plants of the habitat.

teacher may announce that the team that obviously tried harder or had "more difficult choices" than the other is the winner.

As you can see, the element of luck is involved (tossing a larger number on the die) as well as memory and divergent thinking.

Game at a Glance:
1. Teacher writes theme word (e.g., colors) on board.
2. Team A rolls die and gets "3." Team A provides the words "red," "green," and "blue" and gets three points.
3. Team B rolls "1." The team provides the word "orange" and gets one point.
4. Team A rolls "6," but can provide only five color words, so gets five points. (They lose a point for the one word missed.) Team A now has seven points.
5. Team B rolls "4" and provides four words, including the made-up word "Roarange" (combination of red and orange — teacher accepts the word). Team B now has five points.

Possible Theme Words		
ANIMALS	SEASONAL ACTIVITIES	COMPUTERS
PLANTS	CAMPING	BOOKS
CITIES	VACATIONS	TELEVISION
COUNTRIES	TRAVEL	FITNESS
COLORS	FINE ARTS	NUTRITION
STORIES	MATH TERMS	PLANTS
AUTHORS	FRIENDS	FRUITS
FEELINGS	PROBLEM SOLVING	DESSERTS
BEHAVIORS	WILDERNESS	BOY (GIRL) NAMES
SPORTS	HISTORY	FUTURE
CLOTHING	FARMING	TECHNOLOGY
INUIT	CANADA	POLLUTION
PIONEERS	ASTRONOMY	CARTOONS

DISORDER

This non-competitive, cooperative game of cognitive and creative thinking invites players to consider and appreciate correct sequence.

Players:
Groups of five, whole class

Materials:
- Index cards, each giving one step in a sequence of five directions (see "Cards for Disorder" on page 85). Color-coding the card sets, perhaps with dots, so that all five cards relating to a sequence are marked with red, for example, will allow the cards to be easily returned to the correct sets following use.
- Each set of five cards is part of a larger deck, but only one set of five, not presented in sequence, is used at a time.

Rules:
No winner — the goal is to entertain.

Game at a Glance:
The activity "Posting the Letter" (below) has been used here to clarify the steps.
1. One group "performs" at a time; the rest of the class becomes an audience.

Beyond using the line master examples, teachers and students can brainstorm for common nouns that can be described in a variety of ways. For example, while "modem" is quite specific, "parent" can readily be described in five increasingly specific ways and would probably work for the game. This trial-and-error activity, by itself, can be interesting and challenging for students.

This game works well at all ages. Younger kids love the acting. Older students can take it a step further and write the action out in complete, properly sequenced form, taking note of the ambiguous wording and how it could be misinterpreted. Many kids enjoy coming up with action sequences. Disorder is an excellent creative thinking, cognitive activity.

2. The teacher holds one set of five cards like a deck of cards, blank backs facing the students. One at a time, individual group members draw a card.
3. The student conveys what is on the card either by doing charades or reading the phrase aloud. For example, if the student chose "drop it in box," she could act this out.
4. The audience first tries to guess what the small action is. If they guess "mailing a letter," they then try to guess where this step is in the action sequence. They are allowed a single guess only. If their answer is incorrect, the game continues with the next person drawing a card.
5. When the audience is able to identify the overall activity, allow the remaining group members to draw the other cards and present what's on them. The audience still attempts to put all the small actions in sequence.
6. Invite students to guess the exact action title as written on the card.
    Example: (assuming that cards have been drawn in this order)
    Card 1: fold it carefully (the third action in sequence)
    Card 2: stamp it (fourth action)
    Card 3: write carefully and neatly (first action)
    Card 4: drop it in the box (fifth action)
    Card 5: lick and seal (second action)
    ACTION TITLE: Posting the Letter

You can imagine how confusing these actions would be to students who didn't know the correct order or the title. When the actions are presented in this "Disorder," with the words being rather ambiguous, the game becomes both a problem-solving and an "ordering" activity. The creation of a title is a closing activity; only the teacher knows the "true" title.

NAME GAME

This game is either competitive or cooperative, depending on how the teacher wants to use it. A game of reasoning, it facilitates understanding for concise, accurate defining of words, and hence, of problem situations.

Players:
Whole class as two teams (competitive), or individuals or partners (cooperative)

Materials:
- Teacher information on vague-to-specific defining words (see "Possible Name Game Words and Clues" on page 86)
- A team counting list on overhead or board that allows players to see the accumulating points *or* the sheets of paper used by individuals or partners

Rules:
If playing in teams, the team with the highest score wins; if playing as individuals or as partners, individual scores are kept only as personal challenges.

Game at a Glance:
1. Teams determine which team starts, perhaps by playing Rock, Paper, Scissors or doing a coin flip. After the first "call," or guess, teams alternate calling for the rest of the game.

2. The teacher provides the first "clue," a vague, abstract, or indistinct descriptor. See "Possible Name Game Words and Clues." Chances are players will not be able to guess, or "call," the correct name.
3. The teacher provides the next "clue" on the list.
4. Teams alternate "calling" what they think the word is. If a team guesses correctly, that team gets the number of points listed beside the "clue." As clues become more specific, and hence, guessing becomes easier, fewer points are awarded.

Example (based on Team A and Team B)

- Teacher provides the clue word "edible" (worth 5 points).
- Team B has won the coin toss and guesses "pizza" (0 points). (Note: If the team had guessed the word "pomegranate," they'd have earned 5 points, and the teacher would move to the next word.)
- Team A guesses "cake" (0 points).
- Teacher provides the next clue, which is "fruit" (worth 4 points).
- Team B guesses "apple" (0 points).
- Team A guesses "orange" (0 points).
- Teacher provides the next clue, "many edible seeds" (worth 3 points).
- Team B guesses "grapes" (0 points).
- Team A guesses correctly and gets three points.

PRIORITIES

Priorities is a competitive game of problem-solving and of establishing priorities. To some degree, it is also a game of chance.

Players:
Two people or the whole class divided into two teams

Materials:
- Individual Priorities sheets or single transparency for the overhead (see graphic below)
- If using overhead, have a washable pen for reuse of transparency.
- Write problems, one per card, to create a reusable Problems Deck. (See Appendix A, Table of Possible Problems, for ideas.)

Teacher Tip
After playing this game, it is a good idea to discuss how the more succinct, concise, and accurate the clue words, the easier it is to guess the key word. Draw students' attention to how this works when writing or stating a problem to be solved.

Shape of the figure is flexible, but the snake shape seems to appeal to kids. I recommend having at least 10 steps.

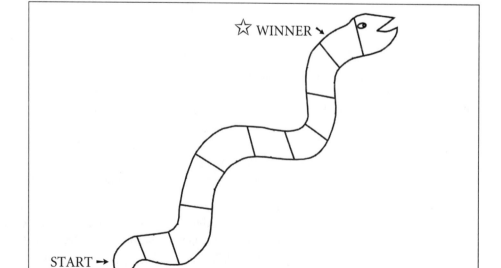

Rules:

First person (or team) to reach the Top Priority wins.

Game at a Glance:

1. All cards are shuffled and spread face down on a table or desk.
2. If playing in partners, each takes one card; if whole class is playing, one player from each team chooses a card in turn.
3. The two players holding cards must decide which card represents the highest priority. This card represents a single move ahead on that player's game sheet. If using an overhead for the whole class, have a different symbol (e.g., ^ or *) for each team, and mark the team's progress with their symbol.
4. If the players cannot agree on which card represents the priority, they present their cases to the teacher (or to a predetermined "judge") who must decide who has the best case and thus will move ahead. If a decision still can't be reached, either no one moves or everyone does.
5. If the players make an obvious error in judgment and choose the wrong card as the priority, the teacher or judge can intervene and have them all move backwards one step.

   Example:
   - First cards exposed: Team A — late for curfew; Team B — have to phone friend

   Late for curfew takes priority so Team A moves ahead one space.
   - Next two cards: Team A — broke my bike; Team B — forgot my homework

(Obviously, kids will have something to say about these two problems. Allow conversation between team members for up to two minutes; then, ask the players who chose the cards to explain why their problems should take priority.)

PROTTO, OR PROBLEM LOTTO

This competitive game involves creative thinking and also lets students see that not only can problems have more than one solution, but that sometimes the same solution works for more than one problem. Many students think that there has to be "just one right way" to solve a problem so this game can be an eye-opener for them.

Players:

Individuals or partners

Materials:

- Individual sheets of paper, pens/pencils
- Visible lists of "problems" to copy (on transparencies or wall charts or handouts that can be collected afterward)
- Possible Solutions Deck: Keeping in mind the problems on the students' list, write possible solutions briefly, one per card. Many solutions will work for several problems. See Appendix B, Table of Possible "General" Solutions.

Rules:

The first student(s) to fill in the sheet according to teacher requirements (e.g., one solution in each quadrant *or* three solutions in three quadrants *or* two solutions in diagonal quadrants) wins the game. The design is flexible.

This game works well for any age, depending on the problems listed. For problems, you can refer to Appendix A, Table of Possible Problems, or invite students to create their own lists and print them legibly on index cards. The latter can be an ongoing project. The students are much better at identifying problems than we are; they know what problems they face.

Remember to refer to Appendix A, Table of Possible Problems, for ideas. Be sure that all problems are worded succinctly as kids tend to copy the shortest ones.

Game at a Glance:

1. Students divide pages into four quadrants by drawing two lines that intersect at right angles and extend to edges.
2. At the top of each quadrant, students write a problem, chosen from the provided list.
3. The teacher randomly draws cards or holds cards with backs facing the students and invites different students to pick a card.
4. The teacher or a student reads the possible solution. If students can fit that solution to one of their chosen problems, they print it in the appropriate quadrant.
5. If a solution fits more than one problem, they can print it in more than one quadrant.
6. The game continues until someone meets the criteria established by the teacher and shouts "Protto," at which point the teacher (or a predetermined panel of student "judges") confirms that the student has met the criteria. Example: One student has selected the following four problems to write in his quadrants:

told a lie	lost my homework
friend wants to shoplift	late for curfew

If the teacher's requirements for this particular game were to "get one solution in each of any two diagonals," this student, with a solution in both the top right and bottom left quadrants, could call "Protto."

- The first solution presented was "do a thorough search." He wrote this correctly in the top right quadrant.
- The next solution presented was "take on a paper route." This idea doesn't fit anywhere on his page, so he could not write it down.
- The third solution presented was "take a deep breath, hold for five seconds, and repeat three times." This solution "step" is appropriate for any problem, so he wrote it in all quadrants (as all other students could have done, too, which means it works against winning).

PULLING POCKETS

This can be a game or an in-class activity. Based on the elements of chance, surprise, and quick thinking, it can be a cooperative game, with teams collaborating to come up with as lengthy a list as possible; a competitive game; or an individual challenge, where each student randomly draws or selects a different pocket — called "pulling a pocket" — and attempts to solve the problem within it.

Players:
Whole class divided into two or more teams approximately equal in size

"Pulling a pocket" is a way to motivate kids and keep their attention since the activity has an element of surprise. It means making a random draw from a choice of "pockets" whose contents — in this case, problems — are hidden. The pockets can be prepared easily by volunteers, aides, or other students.

Materials:
- "Pockets," each filled with a different problem
  A "pocket" is anything that can be closed, such as an envelope, a small tie bag (like the ones used at weddings, available in dollar stores), a small box, an aluminum foil bundle, a newspaper stapled into squares, an empty toilet-paper tube sealed at both ends, a baby food jar, or a plastic egg as found everywhere at Easter. Basically, anything small and hollow works. Each tiny container is filled with a single problem, then closed.

Rules:

The team that "pulls the pocket" answers first. Teams then alternate pocket pulling.

Teams receive points for quickly providing "good" possible solutions. There is an element of subjectivity in determining the excellence of the response, so having a panel of judges who try to be objective (with the teacher's help) is a good idea. Teams alternate giving solutions. They can't repeat a solution already given, and when no one can come up with any more solutions, a new pocket is opened.

Game at a Glance:

1. Students choose pockets and open. If using team approach, first one team chooses a pocket, reads the situation aloud, and then both teams are given 60 seconds to devise solutions.
2. Teams alternately give solutions, earning one point per solution.
3. The team having the most "viable" solutions — offered alternately and not repeated — will have the most points.
4. When there are no more solutions, a new pocket is opened.

   Example:
   - Team A chooses a pocket which contains "lost my cellphone."
   - At start signal, both teams brainstorm solutions to the problem. They stop at signal from the teacher. Signals can be anything familiar to the class, such as a hand clap, a whistle, a flicking of lights, or even just a loud "Stop!"
   - Team A offers first solution. If acceptable, the team gets one point.
   - Team B offers a different solution. If acceptable, the team gets one point.
   - Team A has no more solutions, but Team B has one more, so gets one more point. (Team A has 1 point; Team B has 2.)
   - Team B opens a pocket and continues as before.

SOLUTION BEE

Comparable to a spelling bee, this competitive game encourages students to think broadly, innovatively, and creatively. Solution Bee is a game of luck as well as of divergent thinking.

Players:

Whole class as two teams

Materials:

- Problem–Solution Deck: Use index cards to create this reusable deck; alternatively, you can use paper cut into card-size pieces. The number of cards or papers will depend on the ages of the students, but there should be at least six to begin with. If the teacher finds the game moves perhaps too quickly, more cards can be used the next time. On each card, a problem is identified on one side, with a possible solution written below it. This solution is considered to be the "winning" one, but is not necessarily the "best" one. See page 89. (See Appendixes A and B for more problem and solution ideas; of course, the teacher or students can add more situations.)

There are many other ways to use pockets, too. You can stuff pockets with seasonal greetings, jokes, edibles, actions to perform, words of encouragement, silly fortunes, directions to "prizes," marbles, and statements of appreciation, which students will find much more memorable than verbal comments. Pockets can be reusable or personalized with names and contents specific to each child. Kids love to open their own personal pockets.

Rules:

Teams are required to brainstorm for all the possible solutions to a given problem. They say their solutions alternately, and if theirs is the one written on the card, they win a point. The team with the most points wins.

Game at a Glance:

1. One student from a team selects a card from a fanned deck. The words face the teacher so that there is no way to peek at the problem. The teacher then reads the problem on the chosen card.
2. Teams alternately suggest a possible solution.
3. When a team comes up with the solution on the card, it gets a point; the next problem card is selected randomly by a student and the same procedure is followed.
4. Point out that the solutions on the cards are *not* necessarily the best solutions. Discuss how a single problem can have many solutions, and whether or not the "winning" solution is the best one.

SOLVE-IT-TWICE

This competitive game calls on students to take both adaptive and innovative approaches to problem-solving. It enables players to see that there are at least two equally good, rational problem-solving techniques for a single problem. The assumption is that players know the difference between an adaptive solution and an innovative one.

Players:

Teams — groups, or whole class as two teams

Materials:

- Problems Deck (see page 79)
- Paper and pens/pencils, one for each team
- Overhead, whiteboard, or chalkboard for point counting

Rules:

This game can be judged either by the teacher or by a predetermined panel of student judges. The teams need to come up with two solutions for each problem — an adaptive one and an innovative one. A team receives a single point for a single response, as long as the judge or panel feels that the solution is appropriate. If the team offers a solution that the judges find particularly "amazing," two points can be awarded for that solution. At the end of the game, the team with the most points wins.

Game at a Glance:

1. Each team chooses a player who does the writing and a player who reads aloud what has been written.
2. The teacher or one of the judges randomly draws and reads a card from the Problems Deck.
3. Teams are given 60 seconds (or more, depending on ages and abilities of players) to write down two solutions: one adaptive and one innovative. Talking is allowed between team members.
4. The teacher or a judge calls "stop."
5. Teams present their solutions exactly as written. Judges determine points. The maximum number of points per problem is three: one for the adaptive

solution and one or two for the innovative solution. Teams alternate start-ing.

Example:

- Problem card shows "got caught telling a lie."
- Team 1 writes:
  ADAPTIVE: Apologize. (Judges award 1 point.)
  INNOVATIVE: Apologize, but also show the person a page you've down-loaded about how lying has terrible effects on people and say that you have learned from your mistake. (Judges award 2 points.)
  Total: 3 points.
- Team 2 writes:
  ADAPTIVE: Say sorry. Accept consequences without whining. (Judges award 1 point.)
  INNOVATIVE: Jokingly say that the "devil made me do it." (Judges award 0 points because this solution is not viable.)
  Total: 1 point.

Cards for Disorder

These card steps are in correct sequence. You will need to reproduce them and glue them to cards yourself. The directions are vague for two reasons. If the student chooses to read aloud what's on the card, the words alone will not necessarily give away the final action. If the student chooses to act them out, it can be much more entertaining if he isn't quite sure what he is doing. A lot of guess work is purposefully involved.

ACTION TITLE: BLOWING BUBBLES

1. open the cap

2. pour mixture in a bowl

3. hold the wand carefully

4. blow gently

5. touch one with your finger and watch what happens

ACTION TITLE: WALKING THE DOG

1. attach the hook part

2. open the door, exit, walk, holding the end firmly

3. encourage sitting at the street corner

4. pick up poop and dispose of it

5. continue "controlled" walking

ACTION TITLE: MAKING THE CAKE

1. read the directions

2. gather the ingredients

3. break eggs

4. mix it, pour it, cook it

5. eat and enjoy

ACTION TITLE: MAKING A CHOCOLATE SUNDAE

1. scoop it out into a bowl

2. pour chocolate

3. sprinkle

4. eat and enjoy

5. lick bowl and fingers

Possible Name Game Words and Clues

POMEGRANATE

Edible (5)

Fruit (4)

Many edible seeds (3)

Bright red (2)

Eat seeds only (1)

FOOTBALL

Sport equipment (5)

Team sport (4)

Small and brown (3)

Pigskin (2)

Oval shaped (1)

PENCIL

School tool (5)

Long or short (4)

Filled with graphite (3)

Wooden and pointy (2)

Utensil for writing (1)

TEXTBOOK

Resource (5)

Big or small (4)

Informative, portable (3)

Can be boring (2)

Hardback manuscript (1)

GRAPEFRUIT

Edible (5)

Natural (4)

Fruit (3)

Sphere (2)

Juicy (1)

SODA

Ingestible (5)

Contained (4)

Sweet (3)

Fizzy (2)

Liquid (1)

Sample Problem/Solution Cards

P: I "borrowed" money without asking from Mom's purse and she found out.

S: Gave her my allowance for a month

P: I got grounded on the night of the big game.

S: Try to make a deal with Mom for grounding on a different day.

P: I forgot to pick up little sister from daycare, so they called Dad to come from work to get her

S: Stuck a reminder note to my backpack.

P: When the teacher leaves the room, ___ picks on me.

S: Tell ___ I'm going to beat him up at recess.

P: Class wants to go on an expensive field trip.

S: Have a class bake sale to raise money.

P: Friend wants to cheat from my test.

S: Let her cheat but talk to her later.

P: My best friend told the teacher a big lie and I don't know what to do.

S: Tell the teacher that she lied to her.

P: I missed the school bus and don't have money for a city bus.

S: Call Mom at work.

P: I lost my homework on the bus.

S: Call the bus company when I get home.

P: Friend wants me to shoplift with him.

S: Talk to the school counsellor about it.

Table of Possible Problems

- Accidentally broke a window playing ball
- Accidentally swallowed wrong pills
- Afraid to walk home past a vicious dog
- Afraid of (spiders, ants, bees …)
- Afraid to leave home (for a sleepover, camp trip)
- Am (hungry, thirsty, sick, bleeding …)
- Being chased by a mean dog
- Bike tire flat at school
- Borrowed friend's (coat, sweater, boots) and accidentally ruined (it, them)
- Broke a promise
- Broke your glasses
- Bumped head on monkey bars
- Called someone a bad name
- Came home from school and found mom very sick
- Can't decide which (book, sweater, TV show, movie . . .) to choose
- Climbed on roof and fell off, maybe have broken arm
- Destroyed a library book accidentally
- Didn't do homework
- Didn't keep a secret
- Did something wrong — have to tell parents
- Diplomacy or honesty: Tell a friend something she won't like to hear
- Don't have enough money for a field trip
- Don't want to be mean to a kid but others in your group do
- Drank too much pop and feel sick
- Dropped electronic notebook on the pavement
- Embarrassed by (sibling, parents, relatives, friends, . . .)
- Feel (lonely, bored, worried, sad, . . .)
- Forgot a name which you've been told several times
- Forgot homework
- Forgot mom's birthday
- Forgot to go to dentist appointment
- Forgot to memorize poem to present in class
- Forgot to pick up little sister from school
- Forgot to walk the dog
- Found someone in school bathroom, lying unconscious on the floor
- Friends are vandalizing a car, but you don't want to be involved
- Friend asks to copy an assignment
- Friend asks to copy off you during a test
- Got lost in the city
- Got new clothes dirty
- Given a project partner by teacher — don't get along
- Grounded on party night
- Hit or hurt another student
- Kids you are babysitting are bad
- Lost a letter on way to mailbox
- Lost the neighbor's dog
- Lost your cellphone (Ipod, BlackBerry)
- Mailed a letter without a stamp
- Missed the (game, exam, outing, . . .)
- Missed the bus
- Owe money to . . .
- Relative gave a gift which you gave away — relative coming to visit
- Ripped my good jacket that I wasn't supposed to wear to play
- Seam of pants ripped on way to school
- See a bully hurting a kid, but know that if you tattle, the bully will hurt you
- Skipped (hockey, piano, soccer, dance) practice, on purpose
- Slept through alarm
- Smell smoke in the house, but no one else is home
- Someone wants you to shoplift
- Spilled milk in school lunch room
- The teacher leaves the room and someone begins pushing you
- Think someone is following me
- Told a lie to . . .
- Took money from mom's jar at home
- Want to be on the (soccer, hockey, cheer) team, but don't think you're good enough
- Want to lose weight, but don't want your friends to know you're on a diet
- Witnessed a crime

Table of Possible "General" Solutions

These possible solutions represent some of the most frequently chosen solutions, but they may have to be made more specific for specific problems. For example, "Apologize" may need to be expanded to "Apologize to the homeowner whose window you broke."

- Accept responsibility gracefully.
- Apologize.
- Ask for help.
- Ask how you can make amends.
- Ask if there is a job you can do for money.
- Ask what's wrong.
- Buy glue with your money.
- Call your dad.
- Close eyes and count from 20 backwards.
- Dial 911.
- Do it now.
- Do something else.
- Do something nice for . . .
- Explain to the teacher, without making excuses.
- Explain your feelings.
- Find an alternative.
- Fix (it/them) yourself.
- Ignore (it, him, her) as best as possible.
- Just say no.
- Lie down.
- Look carefully and slowly.
- Look in the library for a helpful resource.

- Make a list of things you can do.
- Offer to get a new one.
- Offer to pay for damages.
- Practice more.
- Rethink everything you did.
- Retrace your steps.
- Say it makes you uncomfortable, so you "can't."
- Say "no," but offer to help in another way.
- Say what you mean; mean what you say.
- Search the Internet for ideas.
- Sit quietly for a few seconds to evaluate the situation.
- Start over again.
- Start saving your allowance.
- Tell a trustworthy adult.
- Tell the truth.
- Tell the truth and accept the consequences.
- Walk away.
- Walk straight home.
- Wait until you are calm.

Unit Plan for Problem-Solving

Unit Anticipatory Set:

- Present a real class problem (e.g., earning money for a field trip; dealing with bullying on the playground). Ask students how they would go about solving the problem. Discuss, then explain that you know a better way to problem-solve. OR
- Invite students to either share with a partner or write in brief about a tough problem they had to solve. If they have trouble starting, disclose a personal (real or imaginary) problem that you couldn't solve.
- Share one of the anecdotes from the book and discuss.

Lesson 1: What Is a Problem?

This lesson is keyed to Chapter 1: Approaching Problems with Care.
Objective: Students will demonstrate familiarity with "problems" through discussion and will understand ways to approach a problem.
Anticipatory Set: Share a poem about a problem (e.g., "Attic Fanatic" by Lori Simmie).

1. Invite responses to these questions: "What is a problem?" and "What problem is represented by the poem?"
2. Teach the seven different approaches to problem-solving. Encourage kids to decide which approach(es) they most often use. Good idea: Create a wall chart with the seven names.
3. Invite discussion about when certain approaches might or might not be used.
4. Follow-up: Have students draw or illustrate in cartoons, any or all of the various approaches.

Lesson 2: Positive Problems and Positive Minds

This lesson is keyed to Chapter 1, especially "1. Creating a Neutral Mind (or, Seeing a Problem Positively)."
Objective: Students will orally demonstrate appreciation of ways in which a problem can be helpful; they will practice achieving a "neutral mind."
Anticipatory Set: Have students "act out" Ostrich, Peacock, and Cuckoo Bird approaches when given the simple problem of picking a partner. **(Ties to previous lesson)**
Share the children's picture book *The Little Engine That Could* by Watty Piper, illustrated by George Hauman. This book leads to the concept of a positive approach.

1. Ask how a problem might be helpful. Discuss the line master "Seeing Problems Positively" (page 18).
2. Follow with a connection to a neutral mind, such as "We've decided that a problem can lead to benefits, so let's now consider what could happen if we looked at it more positively from the beginning."
3. Share ideas for a neutral mind. Have students choose their three favorite ideas and commit them to memory or write about them.
4. Follow up: Ask students to illustrate a "neutral mind" in any way that comes to them by jotting down on paper, drawing, or doodling their thoughts. They may wish to draw a cartoon, paint or color an image of a peaceful place, write about or illustrate symbolically a "relaxed" brain. They may even wish to act out a scenario in which the neutral mind approach is evident.

Lesson 3: How Serious Is My Problem?

This lesson is keyed to Chapter 1, "2. Evaluating a Problem's Seriousness."
Objective: Students will be able to categorize on paper various problems into three levels of gravity and discuss the importance of doing a rapid assessment of a problem's seriousness.
Anticipatory Set: Read the picture book *Chicken Little* (good version by Joan Cusack, Don Knotts, Catherine O'Hara, and Wallace Shawn); doing so will lead to the idea of making a small problem huge.

1. Share a tiny imaginary problem (e.g., a. broken nail) and invite discussion about its seriousness.
2. Share a potentially life-threatening problem (e.g., lost in the forest at night) and discuss how this situation is different.
3. Present the three problem levels on an overhead or board, and have kids fill in two or more "pretend" problems under each heading: Mild, Serious, and Life-Threatening.
4. Discuss how doing a quick calculation of a problem's seriousness is important.
5. Follow up: Students can write a story or play using the Peacock approach.

Lesson 4: Learning the Steps

This lesson is keyed to Chapter 1, "3. Using Rational Problem-Solving Strategies."
Objective: Students will be able to write and remember the five basic steps for rational problem-solving. (At this point they will have no particular understanding of each step.)
Anticipatory Set: Share aloud a poem that involves an order of events, for example, "Dave Dirt," which was selected by Jack Prelutsky for *For Laughing Out Loud*. Ask students to listen to and remember the order of events. Connect the idea of order to sequence and to steps, as in problem-solving steps.

1. Introduce the steps as presented on page 21.
2. Have these steps on a wall chart or individual handouts.
3. Inform students that they will learn about each step separately.
4. Ask students to create illustrated posters representing the Five-Step Problem-Solving Plan.
5. Follow-up: Play the game Priorities (see Chapter 10). This game reinforces the concept of levels of seriousness.

Lesson 5: Getting to Know the Problem, Part 1

This lesson is keyed to Chapter 2: Defining the Problem: Step 1.
Objective: Students will demonstrate orally or in writing the ability to correctly define a problem.
Anticipatory Set: Share copies of the line master "Twelve General Questions to Ask about a Problem" (page 30), or do the sheet together as a class, using an overhead transparency.

1. Give students time to answer the questions based on real or imaginary problems.
2. Review together "Problem-Defining Practices." Work through the five steps together using a possible problem. (Appendix A is always available as a resource.) Take note of the helpful "Teacher Tips" in this section. You may wish to split this lesson into two or even three lessons because it's important that kids get a firm grip on how to word their problems.
3. Follow up: Play the game Protto (see Chapter 10).

Lesson 6: Getting to Know the Problem, Part 2

This lesson is keyed to Chapter 2, especially "Problem Defining Practices" and "A Word about Wording."
Objective: Students will state and write problems properly.
Anticipatory Set: Share aloud a poem that invites discussion about defining a problem. Lori Simmie's "Attic Fanatic," which appears in *For Laughing Out Loud*, is an example. Together, correctly define the problem.

1. Review the previous lesson's concepts.
2. Look together at the line master "Possible Problem-Definition Starts" (see page 29).
3. Supply one or two problems and have students, in pairs, correctly write the problem definitions.
4. If desired, share the ideas offered in "A Word about Wording."
5. Follow up: Play Name Game (see Chapter 10). This game reinforces the concept of concise, accurate wording.

Lesson 7: Collecting Ideas and Possibilities

This lesson is keyed to Chapter 3: Determining Possibilities and Choices: Step 2.
Objective: Students will brainstorm, brain-write, and collect ideas in groups, identifying many possible solutions to a problem.
Anticipatory Set: Use a brainstorming warm-up, such as "The Glad Game" (also found in *3-Minute Motivators*). Partners alternately say, "I'm glad . . . " No repetitions or pauses are allowed. Students keep going for as long as possible.

1. Review "neutral mind" (which is covered in Lesson 2 and discussed in Chapter 1).
2. Discuss brainstorming, brain-writing, inviting ideas, asking a senior, and surfing the Net (but the latter idea is not to be used now).
3. Present a problem and give students about 15 to 20 minutes to generate possible solutions.
4. Discuss by sharing the "best" ideas.
5. Follow up: Play the game Solution Bee (see Chapter 10). This game reinforces the idea that there can be more than one solution to a problem.

Lesson 8: Choosing the Best

This lesson is keyed to Chapter 4: Narrowing and Limiting Choices: Step 3.
Objective: Students will demonstrate how to limit possibilities and prioritize in order to choose the most suitable one(s).
Anticipatory Set: Bring in a selection of "paired sets" (e.g., two different fruits, two different pencils, two different coffee cups). Ask students to decide which is better in each pair. Discuss preferences. Point out that there were few objective reasons for the choices and lead students' attention to making "better" choices by objectively limiting lists of possibilities.

1. Have photocopied lists of many possible solutions for a problem (can come from the previous lesson or from Appendix B, Table of Possible "General" Solutions), one list per pair of students.
2. Pairs then use limiting and prioritizing to come up with two or three "best possibilities."
3. Compare and discuss.

4. As a class, find a way to arrive at a single choice.
5. Follow-up: Play game Solve-It-Twice (see Chapter 10).

Lesson 9: Making It Work

This lesson is keyed to Chapter 5: Acting and Evaluating: Steps 4 and 5.
Objective: Students will create action plans for the previous day's problem/solution, in pairs, and will write one question they will ask to evaluate the effectiveness of the plan.
Anticipatory Set: As a class, play a quick round of "Go-Togethers" (also found in *3-Minute Motivators*). Explain that some things just naturally "go together." Call out a word, and let kids in unison or on their own — not competitive, just fast and fun — call out the "go-together." For example, if you say, "Start," the students may respond with "Finish." Call-out suggestions include *add, multiply, black, day, summer, positive, problem, brother, boys, man, aunt, strawberries, dogs*.

1. Point out how in problem-solving, identifying the best solution is just half of a "go-together" — the other half is the action plan.
2. Discuss action plans.
3. Give students copies of the line master "Action Plan Outline" (page 48), if desired, or allow students to use a linear mode if that's easier. Prompt them to work in pairs or in groups of three to four.
4. Ask how we know if a plan is working. Invite kids to write a single question to evaluate the effectiveness of their plans. (This will be hypothetical.)
5. Follow-up: Share a story or short picture book (*Two Pair of Shoes* by Esther Sanderson is a good one), and ask if the problem presented would require an action plan, and to explain why or why not. Doing this will reinforce the idea that not all problems need the full Five-Step treatment.

Lesson 10: Problem Solved

This lesson is keyed to Chapter 5: Acting and Evaluating: Steps 4 and 5 and Chapter 8: Trouble-Shooting and Consolidating Problem-Solving.
Objective: Students will orally review the concepts learned in the unit and will each begin a Problem-Solving Book.
Anticipatory Set: Share any of the entries in *The Stinky Cheese Man and Other Fairly Stupid Tales*, a wonderful children's book by Jon Scieszka. These tales are all twisted so that the *expected* problem is not there. Read a title and invite ideas about the probable problem. Enjoy the "different problems," and briefly discuss how they might be solved.

1. Introduce the Problem-Solving Books — these can be scribblers, duo-tangs, or anything else available to you. Give one to each child.
2. Daily, each student will date a page and identify a problem from the previous day. Students can do as much or as little as you want, perhaps just identifying the problem or maybe correctly defining it and identifying possible solutions, too.
3. Each student also creates a section, called "My Mistakes," at the back of the scribbler and works forward. Students jot down mistakes they have made and what they have learned, a follow-up for the entire unit.

Classic Fairy Tales with Problems

Many stories lend themselves to a discussion of problems. Although you can no doubt think of a number of such stories, I have identified some familiar tales that could be incorporated into lessons or discussions on problem-solving.

"Rumpelstiltskin," by the Brothers Grimm: A young girl has to find out the true name of a funny little man. She has to solve a problem involving identification.

"The Princess and the Pea," by Hans Christian Andersen: A mother wants to find ways to determine the nature of young women who might wed her son. She has to solve the problem of who is a true princess.

"The Little Match Girl," by Hans Christian Andersen: A young orphan has to find out how to earn money to stay alive. She has to solve a problem of survival.

"The Little Mermaid," by Hans Christian Andersen: A mermaid has to find a way to be with her true love who doesn't live in water as she does. She has to solve a problem of survival.

"Thumbelina," by Hans Christian Andersen: A tiny girl has to find ways to exist in a big world. She has to solve a problem of survival.

"The Frog Prince," by the Brothers Grimm: A frog has to find a way to get a princess to have faith in him. He has to solve a problem of trust.

Index